SKY & TELESCOPE | **Observer's Guides**

METEORS

NEIL BONE

Series editor, Leif J. Robinson

Published in the United States by Sky Publishing Corp.,
49 Bay State Road, Cambridge, MA 02138.

First published in Great Britain in 1993
by George Philip Limited,
an imprint of Reed Consumer Books Limited,
Michelin House, 81 Fulham Road, London SW3 6RB
and Auckland, Melbourne, Singapore and Toronto

Copyright © 1993 Neil M. Bone

ISBN 0-933346-67-0

Edited by John Woodruff
Illustrations by Raymond Turvey
Page design by Jessica Caws

Printed in Hong Kong

Contents

Foreword

A single meteor remains etched in my memory. In the skies over southern California, it literally blew up over my head, producing such a flash that the landscape was instantly transformed into a negative of itself – just as though it had been imaged onto black-and-white film.

And how can I forget the deluge of intriguing reports and photographs that flooded into the *Sky & Telescope* offices after the great Leonid storm of 1966? Processing those accounts allowed me to experience the storm only vicariously, yet it set me planning some thirty years ahead. I'll not miss the Leonids' next great fireworks display if I can help it!

Meteors can be spectacular in the extreme. But such shows are not the stuff of everyday life, or even of everyday science. Each night there are meteors to be routinely logged, mostly from flotsam left over from when our Solar System formed. Yet, about a dozen times a year the Earth encounters a major debris stream from a comet, and observers can then bathe in a meteor shower. These streams are dynamic – they ebb and flow – and that is one reason why continued observation is important.

Neil Bone has provided a great service by preparing a sourcebook in straightforward language for meteor students, casual and dedicated. Just as important, he takes the time to describe in modern terms the meteors' parent bodies and how they came to exist. And, through his great experience, he offers sound advice for observers who want to make a meaningful contribution.

For ill-defined reasons, meteor observing has flagged in North America, in contrast to other regions of the world. This book comes at a propitious time, to revitalize interest before this century ends. With luck, there will be two great meteor displays – in October 1998 the Giacobinids could return with a storm, and only six weeks later the Leonids could do likewise. Bushels of languid Giacobinids and sprinty Leondis in the same year might be too much to hope for, but thanks to Neil Bone we'll be prepared.

Meteor watching can be one of the simplest of all astronomical activities: all you need is pad, pencil, flashlight, and clock. But you also need a very dark sky, and that is becoming harder and harder to find as people spill into every nook and cranny of the world, hell-bent on lighting it up. Perhaps new technologies, such as the charge-coupled device, will get round this problem. Yet, even if it does, there needs to be a long, documented overlap between the traditional visual and photographic observation of meteors and those made by novel means. Along the way, high-tech devices will certainly provide many exciting avenues to explore.

Here is a book to carry meteor aficionados nicely towards the twenty-first century. It provides a wonderful start to a new series of guides for amateurs who wish to explore the unending skies.

Leif J. Robinson
Editor, *Sky & Telescope* magazine

Preface

For all too long, meteors have been the poor relations in the popular astronomical literature. Together with other no less fascinating phenomena of the upper atmosphere such as aurorae and noctilucent clouds, meteors are often dismissed in a couple of paragraphs before the author moves on to the planets or stars. Likewise, there are books on planetary, lunar, variable-star, and deep-sky observing, but there has been no readily available guide to meteors. It is my hope that this book will do something to redress the balance.

Meteor observing is an activity pursued by many amateur astronomers around the world, and one which can be carried out with a minimum of equipment. Naked-eye visual watches can produce scientifically useful data, while more sophisticated equipment – from simple cameras all the way up to low-light video systems and radio receivers – can also be used. It has been my aim to show just what can be done, and how much we still have to learn.

Every year, come August, the dedicated hard core of meteor observers find themselves joined by thousands of "seasonal," once-a-year watchers for the annual Perseid meteor shower, attracting extra interest in 1993 through the return of the shower's parent comet, Swift–Tuttle. We can anticipate a similar, if longer-term upsurge in interest in meteors towards the end of the century, as the possible Leonid storm of 1998 or 1999 approaches.

Perhaps some of the readers of this book will be encouraged to become experienced observers by the time of the major Leonid returns; if so, I will be satisfied that the book has served its purpose. New observers are always needed, and will be welcomed by any of the organizations listed at the end of the book. If my coverage of these topics should appear biased towards the British Astronomical Association's Meteor Section, I can only point to the years of experience accumulated by the BAA, passed on from older hands to newcomers, over the century and more of its existence.

I should like to record my thanks to a number of people who have helped me to put this book together. Dave Gavine and Steve Evans kindly read, and commented on, substantial sections of the draft text. Steve also kindly helped out with photographs of – and from – his rotating-shutter set-up, and of the Perseid spectrum pictured in Chapter 6. Dr Henry Soper promptly answered my plea for some fireball images from his all-sky camera system, and Ron Arbour was similarly quick to provide the photograph of the Orion Nebula used in Chapter 1. John Mason helped greatly with my questions on radio meteor work, and provided Figure 7.3. Special thanks are also due to George Spalding, who, as Director of the BAA Meteor Section from 1980 to 1991, has been a constant source of enthusiasm and encouragement to me, and to scores of other observers. Harold Ridley provided welcome, constructive comments on the first draft of the manuscript, while John Woodruff ensured that errors were kept to a minimum. Last and by no means least, I should thank my wife Gina for putting up with the domestic chaos which attends my writing efforts.

Neil Bone, Director, BAA Meteor Section
Chichester, December 1992

CHAPTER 1

Introduction

What are meteors? Most people have at some time or other seen a "shooting star," the momentary streak of light in the sky which marks the demise of a small particle of interplanetary debris as it dashes through the tenuous upper fringes of the Earth's atmosphere at altitudes of 90 km (55 miles) or so. *Meteors*, to give such events their scientific name, are not exactly rare. At certain times, during some of the major annual *meteor showers*, a trained observer carrying out a systematic watch might record as many

Figure 1.1 *A reasonably bright sporadic meteor, captured on film by the author on 1987 July 29–30, during an 11-minute undriven exposure. (Standard astronomical practice is to give dates in the form "year, month, day"; see also p. 68.)*

as one or two meteors a minute. When there is only minor shower or background *sporadic* activity, the typical haul for a visual observer using only the naked eye may still be between three and eight meteors an hour. The observed meteor rate depends on a number of factors, including not just the time of year, but also the time of night. Estimates suggest that, over the whole world on an average day, there are some 100 million meteors in the naked-eye class.

The incoming particles, called *meteoroids*, which give rise to meteors are mostly rather small – of the same order of size as grains of sand. Meteoroids are travelling at very high speeds relative to the Earth: a meteoroid hits the upper atmosphere at anything between 11 and 72 km/s (between 7 and 45 mile/s), and therefore, despite its small size, carries with it a considerable amount of kinetic energy. This energy is released through collisional processes over just a few tenths of a second to produce a meteor (Figure 1.1). As a general rule, larger particles carry more energy and produce brighter meteors. One the size of a small pebble would produce a very bright meteor, comparable in brightness to the full Moon. Most, however, are much smaller, and give rise to fairly faint meteors.

While similar in size to grains of sand, the bulk of meteoroids entering the Earth's atmosphere lack any of sand's structural integrity. Most of the meteoroids entering the atmosphere are low-density dust fragments which originated in comets. At one time embedded in an icy matrix in their parent comet, these particles have usually long since lost their volatile components (water-ice and gases), and are nothing more than a loose lattice of dusty material. Consequently, cometary meteoroids typically have very low densities, around 0.3 g/cm³. Some astronomers have compared the particles' texture to that of instant coffee granules; others compare their density to that of cork. It is hardly surprising that such insubstantial structures cannot long survive passage through the atmosphere, and there is no known instance of a cometary fragment reaching ground level to be recovered as a meteorite.

Less common are meteoroids believed to have originated from asteroids rather than comets, such as those which produce the Geminid meteor shower in December. These differ from meteors of cometary origin in a number of ways, having a more "rocky" density of around 2 g/cm³, showing less tendency to fragment in flight, and remaining luminous fractionally longer before extinction.

Comets, meteors, meteoroids, and meteorites: some definitions

From the outset, it is important to distinguish between some very similar terms used in meteor astronomy. A *meteor* is the thin streak of light, usually lasting no more than a few tenths of a second, seen in the sky as a "shooting star" or "falling star." The small body that plunges into the atmosphere at high velocity to give rise to a meteor is called a *meteoroid*. Meteoroids come in a range of sizes, from small particles of dust to more substantial chunks of rocky material. There is no hard-and-fast boundary between large meteoroids and small *asteroids*. Most asteroids are small bodies, typically a few tens of kilometres in diameter,

7

Figure 1.2 *One of the brightest comets of the 1980s was IRAS–Araki–Alcock, seen here in a guided 15-minute exposure recorded between 0000 and 0015 UT (see p. 68) on 1983 May 11 by Harold Ridley. The telescope was driven to follow the comet's motion against the background stars, which therefore appear as trails.*

orbiting the Sun between Mars and Jupiter. A relatively small number of asteroids have unusual orbits which cross the Earth's; fragments of these "Earth-crossers" can sometimes enter the atmosphere to become bright meteors. Collisions in the main asteroid belt can also fling fragments into orbits which intersect that of the Earth, and this is believed to be the major source of *meteorites*. Meteorites are (usually!) small pieces of solid material which survive their atmospheric passage to reach the Earth's surface. These are distinct from meteors – it would be wrong for an observer to claim that he or she had seen "a meteorite streaking across the sky last night."

It is also wrong to confuse meteors and *comets*. Contrary to popular misconception, comets do not streak rapidly across the sky. Indeed, unlike meteors, comets are seldom visible to the casual naked-eye observer at all. A bright comet such as IRAS–Araki–Alcock, discovered in 1983, appears as a small fuzzy patch of light moving only slowly against the background stars (Figure 1.2). When at its brightest, IRAS–Araki–Alcock did in fact make unusually rapid progress across the sky, thanks to its proximity to the Earth: over the course of an hour or so, it typically appeared to move about a degree (twice the apparent diameter of the Moon). Even a relatively slow meteor will travel several times this distance in its brief existence. Nevertheless, there are subtle links between comets and meteors. Most of the annual meteor showers appear as the Earth passes through streams of debris released from comets at times when they have passed close to the Sun. These links are discussed in more detail later.

The questions of what meteors are and where they come from have largely been solved by simple observations made with quite basic equipment. Much can be deduced about the nature of meteoroids from careful naked-eye observations and photographs of meteors taken with small cameras. The evolution of the streams of debris in the Solar System which give rise to the regular, annual meteor showers can be studied by comparing systematic visual and photographic observations made by standard methods over many years or decades. Amateur astronomers have collected a vast number of such observations in the past century or so, working through national organizations such as the American Meteor Society and the Meteor Section of the British Astronomical Association.

It is my aim in this book to introduce the simple observational techniques employed by the thousands of amateur astronomers who regularly carry out meteor watches. Meteor observing is an area of astronomical activity ideally suited to the beginner, requiring only the naked eye to be successfully and usefully carried out. At its most basic, meteor observing involves watching an area of sky, and recording certain details of any meteors seen. (This is, of course, an excellent way of getting to know the constellations.) Using binoculars allows fainter meteors to be seen. More sophisticated techniques of photography and even radio observation of meteors are available to those who develop a serious interest in the subject. Work of a standard comparable to that achieved by professional meteor astronomers is possible using readily available modern equipment.

As with any other branch of amateur astronomy, meteor observation should be regarded principally as a source of pleasure, and can be pursued at whatever level suits the observer. I know of several very experienced visual observers who have resisted the challenge of carrying out photographic work in parallel with their watches, for fear of spoiling their simple enjoyment of being out under the stars on a clear night. Many other amateur astronomers restrict their meteor observations to those times when major, dependable showers such as the Perseids or Geminids are active, not wanting to spend the long, character-building blank hours of a cold February night watching for the few meteors visible in the interests of science. I shall show here how useful results can be obtained by carrying out the kinds of simple observation which are a source of great enjoyment to the many amateur astronomers who make them.

Amateur meteor observing

While many amateur astronomers around the world devote much of their observing time to detecting some of the largest structures in the Universe, the distant – and thus often extremely faint – galaxies, meteor observers are looking for the atmospheric effects of material at the other, very small end of the cosmic size scale. Both groups of observers have a common interest in the past.

Time and distance in astronomy

The distances to even the nearest galaxies, such as the Andromeda spiral M31, or M33 in Triangulum, are so large that they are measured in terms of the distance that light (travelling at 300,000 km/s, or 186,000 mile/s) covers in a year. M31 lies 2 million light years away, M33 at a slightly greater distance. Other, fainter galaxies are more distant still. Thus, the deep-sky observer is detecting light which left its source millions, or even hundreds of millions of years ago.

Meteor watching, in a sense, also takes the observer back in time. The particles which are destroyed on atmospheric impact to produce the meteors we see today have, typically, been orbiting the Sun for thousands of years. Tracing their origins further, meteoroids from a cometary source take us back to the earliest days of the Solar System, some 4600 million (4.6 billion) years ago. Comets are believed to be primitive bodies, remaining more or less unchanged since the Solar System began to form. The study of meteors by professional astronomers, augmented by amateur observations, has been an important element in formulating models to explain how the Sun, Earth, and planets came into being.

The origin of the Solar System

For amateur astronomers in the northern hemisphere, autumn nights are among the best for meteor observing. Late October brings the Orionid meteor shower, produced by debris from that most famous of comets, Halley. Watches for Orionids are best carried out in the hours after midnight, as the constellation of Orion, the Hunter, from whose eastern shoulder the meteors appear to emanate, climbs higher into the sky. Orion is one of the best-delineated constellations, with an obvious "belt" of three almost equally bright stars marking his waist. On a good clear night, the early-morning Orionid watcher can hardly fail to notice a compact misty patch, a few degrees south of the belt. This is the famous Orion Nebula, M42, a cloud of gas and dust within our own Galaxy at a distance of about 1600 light years (Figure 1.3). M42 is a stellar birthplace: within its swaddling clouds new stars are forming, stars whose light causes it to shine. Most astronomers believe that the Solar System formed out of a similar cloud.

About 15 percent of the material in our Galaxy is thought to be in the form of such nebulae. These nebulae are, typically, about 25 light years in diameter, and contain the equivalent of a hundred times the mass of the Sun, mainly in the form of hydrogen. Small dust grains, made up of silicates, carbon, and water-ice – probably in a variety of mixtures – are also present. M42 is the most famous example; other prominent nebulae such as the Trifid (M20) and Lagoon (M8), both in Sagittarius, are also thought to be sites of star formation.

Several theories of the Solar System's origin have been proposed, the most popular being nebular contraction. Compression of the pre-solar nebula – perhaps by the shock wave from the explosion of a relatively nearby massive star as a supernova, or the nebula's passage through one of the Galaxy's spiral arms – was probably the initial trigger. Within the large cloud turbulent eddies developed, one of which eventually became the Solar System.

Figure 1.3 *The central region of the Orion Nebula, M42, recorded using a CCD camera by Ron Arbour in early 1992. The Orion Nebula is a stellar birthplace, similar to that which spawned the Sun and planets 5 billion years ago.*

Other eddies in the nebula condensed to form other stars, siblings of our Sun, which have long since gone their separate ways. Observations show that stars tend to form in clusters (also popular targets for deep-sky observers) which gradually disperse as a result of gravitational perturbations. The Pleiades cluster (M45) in Taurus is a good example of a compact, young cluster, containing a couple of hundred hot blue stars and traces of nebulosity.

However triggered, turbulence caused some of the dust grains in the contracting pre-solar nebula to coalesce into small aggregates. These in turn collided with and stuck to other grain-clumps, a process (known as *accretion*) repeated over and over again, until small bodies, or *planetesimals*, became numerous within the eddy. Towards the centre of the contracting cloudlet, the largest collection of colliding planetesimals came together under gravity as the Sun. Such was the mass of material gathered together in the Sun that the pressure of the outer layers on the inner core raised its temperature to the point where nuclear fusion reactions (which continue today, and will do so for at least another 5 billion years) were initiated, and the Sun began to shine.

Dust and other larger particles settled into a relatively flat disk (an *accretion disk*) in the equatorial plane of the Sun. Local condensations of material within this accretion disk developed into the planets, of which Jupiter, having swept up the lion's share of the planetesimals, became the largest. The emerging Jupiter's extensive gravitational pull had a controlling influence on events elsewhere in the pre-Solar System nebula. In a region between Jupiter and Mars lie the orbits of the several thousand 'main belt" asteroids. These are widely accepted to represent planetesimals and fragments of planetesimals in a range of sizes which failed to coalesce into a larger single body as a result of gravitational perturbations by Jupiter. A body about the size of the Earth's Moon would now orbit the Sun between Mars and Jupiter, but for the latter's influence.

The condensation process was quite rapid in cosmological terms, probably taking less than 30 million years. At the end of this time, the strong, gusty particle "wind" blowing outwards from the surface of the young Sun swept away much of the dust lingering in the newly formed Solar System.

Astronomers can observe recently formed stars in a number of regions of the sky, including the Orion Nebula. The first object of this type to be studied in detail was a variable star in Taurus designated T Tauri. Such objects are known as *T Tauri stars* after this protoype, and are characterized by irregular light variations and the presence of associated nebulosity.

Even after the condensation process was largely complete, there remained large numbers of planetesimals. These were gradually swept up by the planets in a late accretion phase of bombardment by asteroid- and comet-sized bodies which ended around 3 billion years ago. The scars of this cratering phase can still be seen on the Moon, Mercury, and Mars, and on the satellites of the outer planets, testimony to a violent epoch in the Solar System's history.

A temperature gradient probably existed within the proto-solar nebula, from its hot interior (where the Sun was forming) to the colder outer regions. This would account for the change from the predominantly rocky nature of the inner (terrestrial) planets to the more gaseous outer (giant) planets. The proto-solar nebula was mainly hydrogen and helium, as well as dusty material and significant traces of the other elements. The early (or *primary*) atmospheres of the planets were derived directly from the nebular gas. During the Sun's T Tauri phase, the terrestrial planets must have lost their primary atmospheres fairly rapidly, since their relatively weak gravitational pulls would have been insufficient to hold on to light gases such as hydrogen and helium in the high-temperature environment of the early inner Solar System. The much more massive giant planets, however, should have retained their primary atmospheres, and future investigation of them by space probes may reveal much about conditions in the proto-solar nebula.

The comets, with their extensive stores of volatile components, are believed to have formed in a broad belt at about the orbital distance of Uranus. Here, the prevailing temperatures favoured condensation of water-ice, and ices of gases such as carbon monoxide (CO) and carbon dioxide (CO_2), together with dusty material, to form small bodies. Main-belt asteroids, with a predominantly rocky composition, are remnants of material which condensed in the warmer inner Solar System. Even across the width of the asteroid belt, there are compositional differences attributable to the temperature gradient in the pre-solar nebula: those which formed in the outer parts of the main belt are generally richer in volatile materials than are asteroids from the inner regions.

Just as gravitational perturbations by Jupiter played a role in preventing the planetesimals in the asteroid belt from collecting together into another planet, so the gravitational influence of the giant planets seems to have dispersed comets from the region around the orbit of Uranus where they formed, into clouds surrounding the Solar System. The main population of the Solar System's comets is believed by most astronomers to reside in a region known as the *Oort Cloud*, a vast reservoir of such bodies. In the simplest models of the comet population associated with the Solar System, the Oort Cloud is a shell surrounding the Sun and extending to a distance of perhaps 2 light years from the Sun – almost halfway to the nearest star.

More recent models suggest the existence of some degree of structure in the comet shell. For example, there may be a sub-population of comets resident in the *Kuiper Belt*, just beyond Pluto's orbit (35–40 times as distant from the Sun as the Earth), and extending outwards to perhaps a thousand times this distance. The Kuiper Belt is proposed by some theorists to be the main source of short-period comets, which are perturbed from it by the gravitational pull of the outer planets. Kuiper Belt comets may differ from those in the main Oort Cloud, in having initially formed in this region of the Solar System, beyond Neptune. These comets are thought to be concentrated mainly in a flat ring lying close to the orbital plane of the planets. The Kuiper Belt, if it exists, may be populated by larger bodies, of which the enigmatic Chiron (classed as an asteroid when discovered, but found in 1989 to have developed a comet-like coma) might be an example. As we shall see later, such bodies could have a role to play in the development of meteor showers.

Astronomers carry out systematic searches for asteroidal bodies. When a suspect asteroid is first found, it is allocated a two-letter designation which indicates the time and order of its discovery. So, for example, 1992 AD was the fourth (D) asteroid discovered in the first half of January (A) 1992; 1992 DA was the first discovered in the second half of February 1992. Later, if further images confirm the object's existence and an orbit is calculated, it is given a number. Well over five thousand asteroids have now been numbered; 1992 AD became asteroid 5145. At this stage the discoverer is usually invited to name the asteroid; 5145 is now also known as Pholus (or 5145 Pholus).

More intensive searches for putative members of the Kuiper Belt, using large telescopes, began to bear fruit in the early 1990s, with the slow-moving, reddish Pholus among the candidates. Another possible Kuiper Belt object, the 200 km (125 mile) diameter 1992 QB_1, was detected as a very faint (magnitude 23) object beyond the orbit of Pluto in autumn 1992. Other such objects doubtless await detection.

Gravitational perturbations by the outer planets would have sent some comets into the inner Solar System, as well as outwards to the Oort Cloud, during its early history. At least some of the impact craters on the planets, the Moon, and other satellites from the late phase of bombardment 3 billion years ago must have been produced by comets.

Another important influence of comets might well have been in the formation or modification of planetary atmospheres, including that of the Earth. The Earth's primary atmosphere was replaced, early in Solar System history, by a chemically reducing atmosphere very different to the one we breathe today. Opinion remains divided as to whether this secondary atmosphere was produced by geological processes such as volcanic outgassing (which continue today, albeit at more modest levels than in the billion years or so following the Earth's formation), or delivered by repeated cometary impacts during the late bombardment phase. Currently accepted biological theories require the early atmosphere of the Earth to have been chemically reducing, containing quantities of gases such as ammonia and methane (which have been found in comet tails by spectroscopy), providing conditions suitable for the chemical reactions which gave rise to life. Life itself has subsequently modified the Earth's atmosphere: the evolution of oxygen-liberating photo-synthesis by green plants around 2 billion years ago set in motion the change to an oxidizing atmosphere.

Comets, then, are believed to reflect the chemical composition of the outer regions of the early Solar System. Slowly orbiting in a near-interstellar deep freeze in the Oort Cloud, those comets which have never passed close to the Sun probably remain largely unmodified from the Solar System's earliest days. Not surprisingly, scientists greet each new arrival from the Oort Cloud with considerable interest: such comets may contain further clues to the origin of the planets, of the Earth's atmosphere, and even of life itself.

Observations of what appear to be dusty accretion disks around stars such as Vega and Beta Pictoris by the Infrared Astronomical Satellite (IRAS) in 1983 have been taken by many astronomers as confirmation that the above scenario for the Solar System's formation is basically correct. It may be that planet formation, as a next step, is a regular accompaniment to starbirth. And it has even been suggested that most stars are surrounded by comet clouds, similar to the Oort Cloud postulated to exist around our Sun. It might even be possible to detect the tell-tale spectral signs of such clouds when their parent stars undergo nova eruptions.

The long and the short of comets

In most years, at least 15 or 20 comets become visible to owners of reasonably large telescopes. Comets are usually visible with such equipment only in the short interval (often a couple of months) during which they are close to *perihelion* – the point in any Solar System body's orbit when it is nearest to the Sun. A small band of dedicated amateur astronomers around the world systematically watches for the appearance of new comets by regularly scanning the sky with "rich field" telescopes or large-aperture binoculars. Patience and persistence are occasionally rewarded by the thrill of discovery, and the added honour of the new comet being named after its discoverer.

Multiple discoveries are common, and it is standard practice for a new comet to be named for those who find it first, up to a third independent observer. Thus, we have such objects as Comet Zanotta–Brewington, Comet Brorsen–Metcalf, and Comet Kobayashi–Berger–Milon. Comet IRAS–Araki–Alcock in 1983 was unusual in having been discovered using a satellite (IRAS), which is given equal credit with two visual observers. The third of these, George Alcock, is a distinguished British amateur astronomer whose remarkable knowledge of the night sky as seen through binoculars has enabled him to make five nova and four comet discoveries. During the 1940s and 1950s, Alcock was also a dedicated meteor observer.

Other prolific comet-hunters of the twentieth century include Bill Bradfield, David Levy, and the huband-and-wife team of the Shoemakers – Carolyn and Eugene. Bradfield, an Australian, has found 16 comets by using the same binocular sweeping technique as Alcock. The Canadian David Levy has gone one better, with 17. Among his discoveries was a splendid bright comet in 1990. The Shoemakers use a Schmidt camera at Palomar Observatory to search for faint objects, and to great effect: their tally of comets now stands at 26, and they look set to break the all-time record.

New discoveries are often first-time intruders into the inner Solar System from the Oort Cloud, and as such are unpredictable. Comets slowly orbiting at the Oort Cloud's vast distance are only weakly bound by the Sun's gravity. As the Sun proceeds in its orbit around the centre of the Galaxy, such comets are subjected to gravitational perturbations by other stars. These perturbations can either cause comets to drift away from the Oort Cloud into interstellar space, or send them sunward.

Many comets have been observed to make their relatively brief dash around the Sun, then retreat back to the depths of the Oort Cloud, not to be seen again, at least on any human timescale. One such *long-period* comet was the infamous Comet Kohoutek of 1973–4. First detected while still very far from the Sun, Kohoutek was expected to be a particularly brilliant object. As it turned out, the comet was rather faint and difficult to see without optical aid, to the disappointment of the public, whose expectations had been raised by articles in the popular press. Kohoutek's long-period orbit takes it out towards the Oort Cloud at *aphelion* (its farthest point from the Sun), and it is not expected to return to the inner Solar System for at least another 30,000 years.

Not all comets seen in a given year are newcomers. Several comets are known whose *short-period* orbits bring them back to perihelion at regular, reasonably predictable intervals. A good example is Comet Encke, which returns to perihelion every 3.3 years, and has been observed on more returns than any other comet. By definition, a short-period comet is one for which the interval between successive perihelia is less than 200 years. Short-period comets are the main source of the streams of debris which produce the annual meteor showers. The system of nomenclature used for comets clearly identifies those with short orbital periods by the prefix "P/," as in P/Encke or P/Halley.

The most famous of all the short-period comets is, of course, Comet P/Halley. In some past epoch, P/Halley, then a member of the Oort Cloud, was dislodged from its stable, distant orbit to fall inwards under the influence of the Sun's gravity. As it neared the Sun on its maiden voyage to the inner Solar System, P/Halley must have been a spectacular sight – a new, fresh comet, rich in reflective dust, a brilliant object in the skies of our far-distant ancestors.

Following its incursion into the inner Solar System, P/Halley – like Kohoutek – would have returned to the depths of space, but for a relatively close brush with the gravitational field of one of the planets. This encounter resulted in the comet's elliptical orbit being modified. Instead of pursuing a long drawn-out orbit, Halley was pulled into a much shorter path around the Sun, with its aphelion a little outside the orbit of Neptune. From tens of thousands of years, Halley's orbital period was reduced to around 76 years. We have no way of knowing precisely when P/Halley was first captured into its shorter-period orbit, but records of the comet are found in historical annals back to 240 BC, and it is believed to have been in its present orbit for some time before this earliest recorded observation.

Encounters with the planets can also have the opposite effect – of flinging comets out of the Solar System altogether, as happened to Comet Lexell in 1779, following a close encounter with Jupiter. Not surprisingly, Jupiter, with its strong gravitational field, is the planet mainly responsible for shepherding comets into short-period orbits. Families of comets, whose orbits reach aphelion around the distance of Jupiter's orbit, have short periods of the order of five or six years.

The predictability of P/Halley's returns led, in 1986, to a small armada of spacecraft being targeted to pass relatively close by its nucleus, and collect valuable scientific data. P/Halley is considered a fairly typical, middle-aged short-period comet, and a reasonable model for other such bodies, and many of astronomers' inferences about comets in general are based on studies of it.

During the early twentieth century, one of the major areas of astronomical debate concerned the nature of cometary nuclei. Opinion was divided between a model which portrayed the nucleus as a "flying sandbank," and the "dirty snowball" model later championed by Fred Whipple. In the former model, the nucleus was held to consist of an aggregation of separate small particles orbiting together. Whipple, however, suggested that a single, solid icy body formed the nucleus of a comet. In either case, it was agreed that the source of all cometary activity, including the sometimes spectacular tails and other phenomena, was the nucleus, responding to heating by the Sun around the time of perihelion. Over time, Whipple's model came to be favoured, and was triumphantly borne out in the images returned on the night of 1986 March 13 by the European Giotto probe, which passed within a few hundred kilometres of P/Halley's nucleus.

Anatomy of a comet

Images returned by Giotto (Figure 1.4) revealed the nucleus of Comet P/Halley to be an irregular body, measuring some 16 by 8 km (10 by 5 miles). When far from the Sun, such a nucleus should be cold and inert, showing none of the activity seen in the inner Solar System. Short-period comets near aphelion are, essentially, indistinguishable from small asteroids in both size and appearance. When, in 1982, P/Halley was first picked up by Earth-based telescopes heading towards its most recent perihelion return, it appeared as an extremely faint, starlike point, out beyond the orbit of Saturn.

A comet's nucleus contains frozen gases and dust, bound together since the earliest days of the Solar System. The outermost layer of a typical comet is thought to be a thin crust, perhaps a couple of metres thick, comprised of dusty material. Beneath this lies a mixture of frozen

Figure 1.4 *The dark nucleus of Comet P/Halley was emitting bright jets of gas and dust at the time of the Giotto probe's closest appraoch (European Space Agency).*

water, gases, and dust. The gas remains frozen while the comet is far from the Sun. Once inside the orbit of Jupiter, however, the nucleus begins to experience more significant levels of solar heating.

One of the Giotto findings which took many astronomers by surprise was the darkness of P/Halley's outer skin. The crust of this comet has an *albedo* (reflectivity) of only 0.03 – as black as soot. (A perfect reflector would have an albedo of 1.00; the Earth's white cloudtops give the planet an albedo of 0.39, making it quite a bright reflector of sunlight.)

Dark surfaces such as that of P/Halley absorb heat efficiently, and a comet entering, or returning to, the inner Solar System therefore warms up quite rapidly. As the temperature rises, pockets of frozen gas beneath the surface begin to sublime – passing from the solid state straight to the gaseous state – and the expanding gas then begins to emerge, sometimes explosively as jets, through cracks in the comet's thin outer crust. Another important Giotto finding was that the gas-emitting areas on P/Halley's nucleus were fairly limited, covering only 10 percent or so of its surface. "Hot spots," made active by solar heating, appear to be the source of persistent gas jets in cometary nuclei.

Initially, the gas emitted in these jets stays fairly close to the nucleus. Excitation by the ultraviolet component of sunlight causes the gas to fluoresce, glowing as a small *coma* around the nucleus. Atoms and molecules of gas in the coma become stripped of their electrons – *ionized* – by the continued action of solar radiation. Even the highest-resolution Earth-based images show nothing more than a comet's coma; the actual nucleus remains hidden within the surrounding gas cloud.

The entire Solar System is permeated by a continual, sometimes gusty outflow of ionized gas from the outermost regions of the Sun's atmosphere, the *solar wind*. This wind flows past the Earth at an average

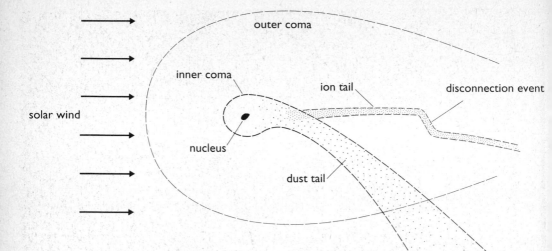

Figure 1.5 *A schematic representation of the principal features of a comet. The source of the comet's activity, the nucleus, is invisible through Earth-based telescopes, being lost in the surrounding light of the coma. Dusty material falls away into a curved tail, while ions stream out to form a straight tail which may alter its orientation in response to changes in the solar wind flowing past the comet.*

velocity of 400 km/s (250 mile/s), carrying with it an *interplanetary magnetic field*. As they pass a comet, the lines of force of this magnetic field become "draped" – rather like the spokes of an umbrella – around the cometary nucleus. Cometary ions from the coma are picked up by these field lines, and guided into the *plasma* or *ion tail* (Figure 1.5).

Comets' ion tails are usually fairly straight and point more or less directly away from the Sun, but can vary to some degree in their orientation, depending on the local interplanetary magnetic field carried along in the solar wind. Ion tails can extend for millions of kilometres across the Solar System, downwind of an active comet. Sometimes, if the magnetic field's direction changes abruptly, an ion tail can be sheared off in a *disconnection event*, to be replaced over several hours by a new tail growing in a different direction. Several such events were seen during the 1985–6 apparition of P/Halley, which coincided with an unusually vigorous period of solar activity near the end of the sunspot cycle. Ion tails often show a bluish colour, characteristic of emissions from excited carbon monoxide gas.

Many active comets show a second type of tail, produced by dust liberated from the nucleus as gas sublimes away. In particular, jets of gas will carry away small fragments of the surrounding crust, which fall away on either side of the comet, and spread out into broad, curved or fan-shaped *dust tails* (Figure 1.5). A particularly fine example of a comet showing both types of tail was Comet Bennett in 1970 (Figure 1.6).

Dust tails often show a more yellowish colour, typical of the *reflection* of sunlight by small particles, in contrast to the bluish *emission* stimulated by

Figure 1.6 *Comet Bennett, one of the finest comets of the past fifty years, clearly showing the contrast between ion and dust tails. This 40-minute exposure early on the morning of 1970 April 4 was obtained by British comet observer Michael J. Hendrie. As with Figure 1.2, the telescope was driven to follow the comet rather than the background stars.*

short-wavelength solar radiation in ion tails. The particles which produce the dust tail are generally very small – smaller, on the whole, than those which give rise to meteors in the Earth's atmosphere. The larger particles, which are of interest to us in meteor studies, are more massive, and do not escape quite so far from the nucleus, at first. Gradually, however, they do spread out to produce *meteor streams*, some of which produce meteor showers if their orbit and that of the Earth happen to intersect. Before going on to discuss the production and evolution of such streams, it will be useful to take a look at the orbital characteristics of comets (and of the particles liberated from comets), and how these are defined. The development of meteor streams is directly influenced by how particle orbits become altered by the long-term gravitational effects of the planets.

Orbits

The orbital characteristics of comets, and of other Solar System bodies around the Sun may be defined by a number of parameters, as I shall describe shortly. Our understanding of orbital mechanics is the product of centuries of observation and research. The seventeenth-century mathematicians Isaac Newton and Johannes Kepler determined the basic laws required to provide an adequate explanation of the movements of the planets. Kepler's work, based on the meticulous positional measurements made by his mentor and benefactor, the Danish astronomer Tycho Brahe, is summarized in his three laws of planetary motion. Kepler's laws apply not only to the planets, but also to comets and asteroids, and even meteoroids.

Kepler's first law states that the orbit of each planet is an ellipse, with the Sun at one of the two foci (Figure 1.7). As we have already seen for comets, a consequence of the elliptical shape of a body's orbit is its arrival at points closest to (perihelion) and farthest from (aphelion) the Sun at certain times. The precise shape of an elliptical orbit is given by its *eccentricity* (*e*), the deviation of the ellipse from perfect circularity. Geometrically, the eccentricity of an ellipse is given by the distance between its two foci, divided by the length of its major (longer) axis. For a perfect circle the foci coincide, and the eccentricity is zero. The Earth's orbit around the Sun is not too far from circular, having an eccentricity of 0.0167. Comet P/Halley, on the other hand, has a markedly elliptical orbit with an eccentricity of 0.967.

Other characteristics of a Solar System orbit may be understood by reference to defined coordinate systems widely used in astronomy. In stating the positions of astronomical bodies as observed from the Earth, it is useful to employ the concept of the *celestial sphere* (Figure 1.8). This may be envisioned as a shell surrounding the Earth at a great distance, onto which the stars and planets are studded – rather as they appear when simulated by projection onto the dome of a planetarium. Onto the celestial sphere is projected a coordinate system of *right ascension* (RA) and *declination* (dec), equivalent, respectively, to latitude and longitude on the Earth. Projected onto the celestial sphere, the Earth's equator becomes the *celestial equator* which, like its terrestrial counterpart, is a great circle (one which divides the sphere into two equal hemispheres).

As the Earth orbits the Sun, the Sun appears to trace out a path around the celestial sphere, moving gradually eastwards through the constellations of the zodiac. This movement led the ancients to the (incorrect) assumption that the Sun, and everything else in the Solar System, revolves around the Earth. We know differently nowadays, but the apparent path of the Sun around the Earth's sky usefully traces out a second celestial great circle of importance in understanding the motions of Solar System bodies, the *ecliptic*. Put simply, the ecliptic marks the plane of the Earth's orbit. The orbits of all the major planets

Figure 1.7 *The orbit of a comet, like that of any other body in the Solar System, is an ellipse, with the Sun at one of the two foci. At its closest to the Sun, the comet is said to be at perihelion; at its most distant it is at aphelion.*

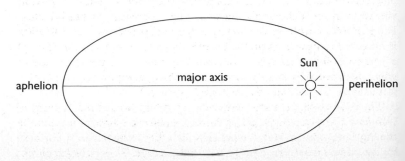

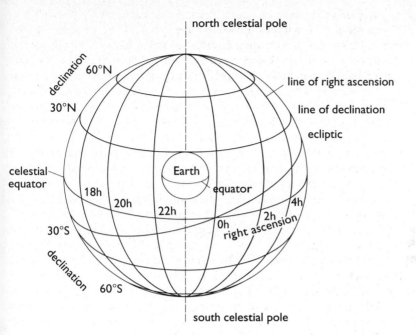

Figure 1.8 *A convenient way of describing positions in the sky: the artificial concept of the celestial sphere, seen here from the outside. An observer on the Earth, at the sphere's centre, sees only one hemisphere at a time. Celestial bodies have positions defined in right ascension (RA) and declination (dec) — corresponding, respectively, to terrestrial longitude and latitude.*

lie fairly close to the same plane, the plane passing through the Sun's equator (an observation which leads to the inference that all were formed in a fairly flat accretion disk, as discussed earlier).

The tilt of a planet's, or any other Solar System body's, orbit relative to the ecliptic is termed its *inclination* (*i*). (This and other *orbital elements* are shown in Figure 1.9.) So, the orbits of the major planets (not counting Pluto) have low inclinations, lying within 7° of the ecliptic. Several comets and some asteroids, however, have highly inclined orbits. Comet P/Halley has an orbit effectively inclined at 17.8° to the ecliptic, and spends much of its time well below the ecliptic plane. Cometary orbits may be *prograde* (in the same direction as that of the planets – anticlockwise as viewed from above the north pole of the Sun), or *retrograde*, like P/Halley. From Earth, an object in a high-inclination orbit, such as P/Halley, is observed to pass through the ecliptic plane in an upward direction at a certain point in its orbit, the *ascending node* (☊), the longitude of which is denoted by Ω. Similarly, as the object dives below the ecliptic, it is said to pass through the *descending node* (☋).

The position of the Earth in its own orbit may be defined in terms of *solar longitude*, given by the Sun's position along the ecliptic. As a result of the Earth's 23.5° axial tilt, the plane of the ecliptic appears to be inclined by the same angle relative to the celestial equator. Thus, in its apparent

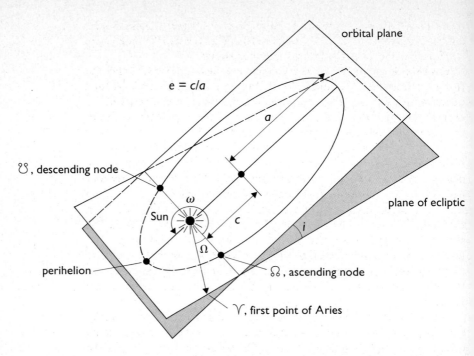

Figure 1.9 *Orbital elements: inclination (i), longitude of the ascending node (Ω), argument of perihelion (ω), semimajor axis — half the ellipse's greatest diameter (a), and eccentricity (e). The eccentricity is given by e = c/a, where c is the distance from the centre of the ellipse to one of its foci. Together with the period (P) and the time of perihelion passage (T), these elements completely determine an orbit.*

motion around the ecliptic, the Sun crosses the celestial equator going north at one point, and crosses it going south at another point six months later. These points are the *equinoxes*. For northern hemisphere observers, the northward crossing around March 21 marks the spring or *vernal equinox*, while the Sun's southward passage across the celestial equator is at the *autumnal equinox*. The intersection of the ecliptic and celestial equator at the position of the northern hemisphere vernal equinox marks the zero point of solar longitude, and also the zero point of right ascension. It is known as the *first point of Aries*, denoted by the symbol ♈.

The apparent ecliptic longitude of the Sun from the first point of Aries increases by about one degree per day. The precise amount of daily increase depends on the time of year. From Kepler's third law of planetary motion, we find that when a planet is near perihelion its orbital movement is faster. Hence solar longitude appears to increase most rapidly during December and January, when the Earth is near its perihelion.

Meteor streams whose orbits' ascending or descending nodes relative to the ecliptic lie close to the Earth's orbit will be encountered at the same (or very nearly the same) solar longitude from one year to the next. Subtle changes in the solar longitude at which maximum activity is seen, revealed by careful observations, may provide clues to the evolution of meteor streams, or reveal that gravitational perturbations by the major planets are physically shifting a stream's orbit. These aspects are considered in a little more detail in Chapter 5, where we look at the main annual meteor showers and their behaviour.

The ecliptic longitudes of its ascending and descending nodes help to further define the three-dimensional configuration of an orbit. A final quantity, the *argument of perihelion* (ω), indicates the angle between the orbit's ascending node and perihelion as measured from the centre of the Sun, in the direction of the comet's motion.

Another essential orbital element is some indication of the distance scale. It is convenient, in most discussions of Solar System orbits, to use distances defined in astronomical units (AU). One AU is equivalent to 149,597,870 km (92,955,730 miles), the mean orbital distance of the Earth from the Sun. For instance, the orbit of Jupiter lies 5 AU from the Sun, while that of Saturn is at the greater distance of 9.5 AU. The scale of an elliptical orbit is given by the value of the *semimajor axis* (a), defined as half the distance along the ellipse's greatest diameter (the major axis), and expressed in AU. The semimajor axis is equivalent to the average distance of the orbiting body from the Sun. For Comet P/Halley, $a = 17.939$ AU.

In our consideration of comet and meteoroid orbits, it is important to know the distances at which a number of key orbital elements occur. If an orbit's ascending or descending node relative to the ecliptic is close to 1 AU from the Sun, then it will be intersected by the Earth's orbit. For comets, the distance from the Sun at perihelion affects the level of activity in the nucleus, and therefore how much dust is liberated. Comet P/Halley comes to perihelion well within the Earth's orbit, at a distance of 0.59 AU, and is subjected to considerable solar heating around this time. For short-period comets, the anticipated time at which perihelion will be reached (T) can be predicted reasonably accurately beforehand, and we have a good idea of the period (P) between successive perihelia.

Putting all these orbital elements together, we can describe the orbit of Comet P/Halley at its last return as follows:

T 1986 Feb 9.43867
ω 111.84658°
Ω 58.14397°
i 162.23932°
e 0.9672725
a 17.9390115 AU
P 75.98 years

For comparison, the orbit of a typical meteoroid in the Orionid stream has the following elements:

ω 82.5°
Ω 28.0°
i 163.9°
e 0.962
a 15.1 AU

The similarity of the values of i and e, particularly, imply a link between Comet P/Halley and the Orionid meteor stream.

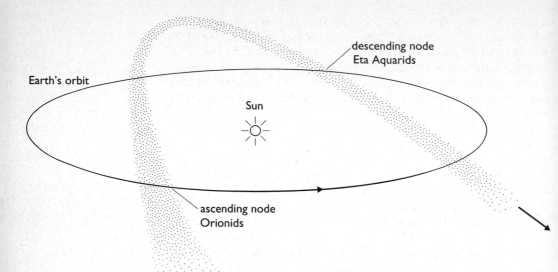

Figure 1.10 *Meteor showers result from the Earth's passage through streams of cometary debris. The P/Halley debris stream is encountered twice: at its descending node in May, to produce the Eta Aquarids, and at its ascending node in October, producing the Orionids.*

The Earth's passage through the P/Halley stream is shown in Figure 1.10. Similar associations between other comets and meteor streams have been found. The Perseids, for example, are associated with Comet P/Swift–Tuttle, seen in 1862 and again in 1992–3, while the Taurid meteor stream is debris from Comet P/Encke. Given an understanding of these orbital connections, we can now look in more detail at how meteor streams come into existence, and how they evolve with time.

The evolution of meteor streams

Nearly all the meteors that are observed result from the entry into our atmosphere of particles which originated in comets. These particles are liberated around the time of their parent comet's arrival at perihelion. The Giotto probe found that the nucleus of Comet P/Halley was pouring forth as much as 3.1 tonnes of dust every second, a week or so after its 1986 perihelion. Gas jets emerging from below the surface carry away fragments of the comet's crust and other dusty material into the coma surrounding the nucleus.

The dust particles travel outwards from the nucleus along paths which are initially fairly straight. However, the solar radiation pressure has a significant influence after a time, pushing the smaller particles out into orbits which differ from that of the nucleus. The curvature of cometary dust tails results from a combination of radial and orbital motion; meteoroids in the dust tail lag progressively farther behind with increasing distance from the nucleus.

The heavier fragments – typically of the order of 0.1 g in mass – escaping from the nucleus eventually end up pursuing independent orbits around the Sun as members of a meteor stream. Such streams take some time to develop, and closed loops of debris such as that which produces the Perseid meteor shower each August represent a mid-point in meteor stream evolution. The Perseids are an excellent example of a reasonably well-evolved stream; the Earth's motion around the Sun brings the opportunity to observe meteor streams at earlier and later stages of development.

Initially, the dust which will go on to form a meteor stream stays fairly close to the nucleus, returning to perihelion at more or less the same time as the parent comet. Successive perihelion returns add more and more dust to the nascent meteor stream, forming a cloud of meteoroids – a *meteor swarm* – in the vicinity of the comet's nucleus. Such very young streams produce significant levels of meteor activity only in those years when the comet is close to perihelion. Encounters between the Earth and the meteor swarm lagging behind Comet P/Tempel–Tuttle have given rise to short-lived, very intense bursts of Leonid meteor activity – *meteor storms* – seen at 32–33 year intervals under favourable conditions. Leonid activity in the years when the comet and its attendant meteor swarm are at aphelion is much lower, but not, as many popular texts would have it, altogether absent. Observations of the pattern of Leonid activity in these quieter years give some indication of how debris released at past perihelia is spreading out around the orbit. The importance of such observations will be discussed further in Chapter 5.

Several effects combine to spread meteoroids from a swarm into a complete loop around the stream's orbit. The initial ejection of dust is more or less random in direction: as the comet's nucleus rotates, its gas jets spray dust particles continually so that some particles are ejected ahead of the comet, and some behind. Once ejected, each meteoroid pursues its own elliptical orbit around the Sun. Particles ejected ahead of the comet will arrive at perihelion before the comet does, while those ejected behind will reach perihelion afterwards.

While in orbit around the Sun, each meteoroid is subjected, like all other Solar System bodies, to gravitational perturbations. As might be expected, these perturbations are much more marked for low-mass meteoroids than they are for the cometary nucleus from which they are ejected. Consequently, meteoroids ejected ahead of or behind the comet will, respectively, arrive at perihelion progressively earlier or later than (and farther from) the nucleus. Over time, particles from behind the comet meet those from ahead of the comet, and the meteor stream becomes a closed loop, as shown in Figure 1.11.

When the Earth runs through such a loop of debris, we see a meteor shower. Shower meteors can be relatively easily identified. Since the meteoroids in a particular stream have parallel orbital paths, they also have parallel trajectories into the Earth's atmosphere. A result of this is that meteors from a particular shower appear to emanate from a particular direction in the sky, the *radiant* (Figure 1.12). The radiant effect is a result of perspective. Just as the parallel sides of a straight road appear to converge at the horizon, so meteors from a common source seen over a large portion of the sky appear to follow paths diverging from the same point.

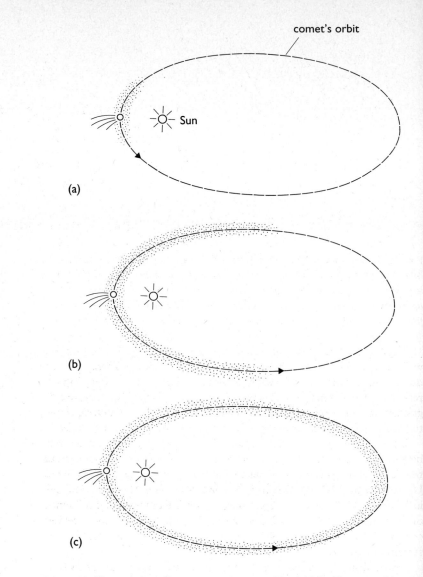

comet's orbit

Sun

(a)

(b)

(c)

Figure 1.11 *The evolution of a meteor stream over time. Initially (a), meteoroids are in a swarm close to the parent comet. Gradually (b) the swarm spreads ahead of and behind the comet until, eventually, meteoroids trailing the comet meet those preceding it to complete the loop (c).*

The Perseids, from Comet P/Swift–Tuttle, and the Orionids, from Comet P/Halley, are good examples of streams at this second stage of development. Activity is relatively consistent from one year to the next. The Perseids have a reputation as one of the most dependable showers: activity lasts for a couple of weeks each August, providing an annual highlight for meteor observers. Occasional strong bursts of activity, as in 1980 and 1991, probably result from encounters between the Earth and denser filaments of debris within the stream.

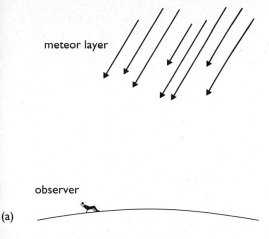

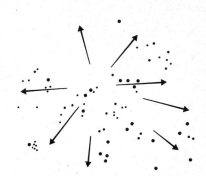

Figure 1.12 *The radiant effect allows observers to identify the stream a particular meteor might belong to. Meteoroids from a common source travel around the Sun in parallel orbits, and any resulting meteors will also have parallel trajectories in the upper atmosphere (a). Perspective, however, causes such meteors to appear to diverge from a single area of sky – the radiant – as seen, for example, during the Perseids (b).*

There is evidence that filaments of debris within the Orionid stream were released from P/Halley's nucleus at different epochs. Filaments laid down at times when the comet's orbit was in a slightly different inclination result in several separate sub-peaks in Orionid activity. In response to gravitational perturbations, the comet's orbit "nods" up and down relative to the ecliptic, and the ejected dust has been well spread out as a result.

With the greater availability of more advanced computers, researchers have been able to model the evolution of meteor streams. Prominent in this field have been research teams led by David Hughes at Sheffield University, and Iwan Williams at Queen Mary and Westfield College in London. Their results suggest that, under the gravitational influence of the planets, the 7-year orbit of one of Jupiter's family of short-period comets should take about 300 years to become filled by a continuous stream of meteoroids.

Gravitational perturbations by the planets do much to smear out meteor streams. Particularly around aphelion, meteoroids can become pulled into modified orbits – or even lost from the stream's orbit altogether – with the result that an initially compact stream (like the Perseids) becomes more diffuse with time.

The scatter of meteoroid orbit aphelia is well illustrated by the Quadrantids. Reflecting their release mainly around perihelion, Quadrantid meteoroids return to perihelion in a fairly tight zone, around the same ecliptic longitude. Thus, when encountered by the Earth the stream is narrow, producing a short, sharp peak in activity in early January lasting only about 6–8 hours. By contrast, the aphelia of Quadrantid meteoroids are so widely scattered in distance from the Sun, and in longitude, that it takes Jupiter about 4 years of its 12-year orbit to pass from one edge of the stream to the other.

Other effects can lead to the loss of particles from meteor streams. The best known of these is the *Poynting–Robertson effect*, whereby the smallest particles are lost preferentially. The Poynting–Robertson effect is a consequence of meteoroids' absorption and subsequent re-radiation of solar radiation. In re-radiating the energy received from sunlight, meteoroids lose some of their orbital velocity, and spiral gradually in towards the Sun. The effect is most marked for small particles, which have the largest ratio of surface area to volume.

A consequence of the long-term operation of the Poynting–Robertson effect is that an old meteor stream becomes depleted in smaller particles. The brightness of a meteor is determined by two principal factors – relative entry speed and size of the incoming meteoroid. Size for size, the faster a meteoroid enters the atmosphere, the brighter it will be. Likewise, speed for speed, the larger (and, usually, more massive) a meteoroid, the brighter the resulting meteor. Examination of observers' brightness estimates (I shall show how these are obtained in Chapter 4) gives some idea of the distribution of particle sizes in a stream. Older streams tend to produce proportionately fewer faint meteors than young streams. As a middle-aged meteor stream, the Perseids show a slight depletion in the observed proportion of faint meteors. The effect is more marked for the Taurid meteor stream, left in the wake of Comet P/Encke.

The Taurids are an excellent example of a meteor stream approaching old age. Not only has the stream lost many of its smaller particles, it has also had sufficient time to become spread out over a great swathe of the inner Solar System by the gravitational influence of the planets. Consequently, the Earth spends a couple of months traversing the Taurid stream, which produces fairly low observed meteor rates during October and November.

The ultimate fate of an old meteor stream will be for its rates to become so low that it no longer produces a noticeable shower. At this stage, its activity will have merged with the sporadic meteor background produced by the dust which fills the inner Solar System.

Giant comets, comet "showers," asteroids, and mass extinctions

The discovery in 1977 by Charles Kowal, working at the Palomar Observatory, of the apparently asteroidal object 2060 Chiron raised some interesting questions about the distribution and evolution of small bodies in the Solar System. Chiron was detected by its slight movement against the background stars on a "deep" photographic plate exposed to record faint images. While it produced a trailed image, as expected of an asteroid, Chiron was clearly unusual. Its trail was much shorter than would be expected for a member of the main asteroid belt between Mars and Jupiter. Once a few more images had been obtained for precise positional measurement, Chiron's orbit could be determined, and it was found to occupy an unstable orbit between Saturn and Uranus.

Chiron was discovered ahead of its perihelion, which is at 8.5 AU. Continued observations showed it to be brightening slowly, and on the basis of typical asteroid albedos, the amount of light it reflected was taken to indicate a diameter of 250 km (150 miles). Chiron had a further surprise in store. In 1988, observations from Kitt Peak National Observatory in Arizona revealed the object to have developed a coma. Gas emission from Chiron led to an increase in its brightness and apparent diameter – behaviour more typical of a comet.

It is now thought likely that Chiron is a giant comet, perhaps captured from the proposed Kuiper Belt, in an unstable orbit. At some time in the next 10,000 years, a close encounter with Saturn will fling Chiron from its current orbit, possibly into the inner Solar System. The consequences of such an event have been discussed by many researchers, including Victor Clube and Bill Napier at Oxford University. Entering the inner Solar System, a giant comet would be subject to the same heating as its smaller Oort Cloud brethren. Fragmentation could occur as stresses built up under the crust. Several smaller comets have been seen to break up close to perihelion – as happened, for example, to Comet West in 1976, whose nucleus fragmented into at least four parts.

Clube and Napier have proposed that the Taurid meteor stream was produced in the far distant past by just such a mechanism. Disintegration of the Taurid parent body allegedly produced Comet P/Encke, and a number of other smaller bodies (both comets and asteroids), together with copious amounts of dust, all sharing a common orbit. Their theory suggests that we are seeing this complex at a late stage in its evolutionary development. Thousands of years ago the Taurid complex may have produced considerably more meteor activity, as it encountered the Earth, than at present.

Pockets of higher meteoroid concentration may still exist within the stream, and it has been suggested that the stream includes small asteroids. The object which gave rise to the Tunguska event, in which a huge explosion devastated a large area of the remote Siberian tundra on 1908 June 30, might even have been an example of the latter class of object. The blast occurred during the time of year when material from the Taurid stream produces the daytime Beta Taurids, which have been detected by radio techniques.

The airburst which produced the Tunguska explosion was spectacular, and its consequences would undoubtedly have been yet more devastating had it happened above a major city. Napier and Clube are among many astronomers who have suggested that past collisions between the Earth and bodies even larger than the putative Taurid fragment which caused the Tunguska explosion have had extremely far-reaching consequences.

A certain population of asteroids have orbits which cross the Earth's. In theory, something like ten thousand such *Apollo asteroids* might exist. Apollo asteroids could be the nuclei of comets which are no longer active in producing gas or dust, or bodies flung into Earth-crossing orbits by collisions in the main belt, which lies between Mars and Jupiter. Many Apollo asteroids are very small – for example, the object 1991 BA, which passed the Earth at a distance of 170,000 km (105,000 miles), a very near miss in Solar System terms, was about 10 metres

(30 feet) across. None the less, such a body would produce a crater at least ten times its own diameter on impact.

There is a small but definite chance, over very long timescales, of collision between the Earth and a much larger Apollo asteroid. In many theories a collision between the Earth and a relatively large (10 km/ 6 mile diameter) Apollo asteroid hastened the demise of the dinosaurs in a mass extinction 65 million years ago. The impact of such a body could, theoretically, have brought about temporary changes in the climate, destroying the ecological systems on which the dinosaurs depended. Possible evidence for such an event is the presence of an iridium-enriched layer in the sedimentary rocks laid down around this time. Iridium is a rare element in the Earth's crust, but is more abundant in meteorites, which are believed to be mainly asteroidal in origin. Researchers believe that they may have found the impact site, at Chicxulub, in the Yucatán Peninsula in Mexico.

An alternative cosmological extinction theory invokes comet "showers" – interludes during which larger numbers of comets are more frequently ejected sunward from the Oort Cloud. A collision between the Earth and one of these bodies would then be likely, with the same consequences as for an asteroid impact. The question of whether the apparent mass extinctions at the geologists' Cretaceous–Tertiary boundary 65 million years ago were caused by a cosmic impact is one which is difficult to answer, but it has led to much lively debate among not just astronomers, but also geologists and biologists.

The zodiacal dust cloud

As the Earth travels in its orbit around the Sun, it encounters meteor streams at various stages of their evolution. The Leonids, at one extreme, are largely concentrated in a compact particle cloud near their parent comet. The Perseids are a middle-aged stream, still fairly compact but distributed more or less evenly around the orbital ellipse out to the distance of Saturn. Less readily recognizable, but actually one of the two most significant sources of meteoroidal dust in the inner Solar System, are the Taurids, which have become so spread out that they are encountered by all four terrestrial planets – Mercury, Venus, Earth, and Mars.

The inner Solar System is laced by unknown numbers of such streams: we can only be aware of those few which are encountered by the Earth. Others lie in orbits whose ascending and descending nodes pass nowhere near the Earth, at least at the moment. Gravitational perturbations will no doubt bring some of these showers "on stream" at some future epoch. At least one relatively modern shower, the Bielids (also called the Andromedids), is currently in a modified orbit such that it no longer encounters the Earth. It appears from computer simulations of the stream's orbit carried out by David Hughes at Sheffield University that this shower, which produced spectacular storms in 1872 and 1885 but was later posted missing, presumed dead, will reappear in the Earth's skies around the year 2120.

In some respects meteor streams may be regarded as denser strands within the vast dust cloud which fills the inner Solar System out to the orbit of Jupiter, at a distance of 5 AU from the Sun. Because it lies in the plane of the ecliptic, the path the planets appear to follow as they move against the background constellations of the zodiac, this cloud is known as the *zodiacal dust cloud*. The dust has its origins in the gradual decay of meteor streams, themselves the decay products of short-period comets. It has been estimated that the zodiacal dust cloud contains a mass of material equivalent to a single comet, and that without continual replenishment by dust from fresh comets reaching perihelion passage it would disappear in a few thousand years.

Under ideal conditions (and, sadly, rather rarely from locations suffering from artificial light pollution), the reflection of sunlight from this dust may become apparent as the *zodiacal light* (also known as the "false dawn"), a cone of faint luminosity comparable to the Milky Way in brightness. Best seen from the tropics, the zodiacal light may occasionally be glimpsed from temperate latitudes during autumn mornings or spring evenings. Even more difficult to detect is the pale glow of the *gegenschein* (or counterglow), which appears directly opposite the Sun in the sky. The *gegenschein* is also produced by the reflection of sunlight from zodiacal dust. This light is spread out over an oval region covering 10–20° of the sky, and is therefore very diffuse, and tends to be lost against the general light-polluted background. The *gegenschein* and zodiacal light can, under exceptional conditions, be seen to be connected by a faint *zodiacal band*.

The zodiacal dust cloud was studied by the IRAS satellite in 1983; the dust absorbs solar radiation, and re-emits significant amounts of infrared. IRAS observations revealed bands of dust within the asteroid belt, presumably resulting from collisions between these bodies. As one professional astronomer has put it, the asteroid belt is gradually grinding itself down. Other spacecraft investigations have included the use of photopolarimeters aboard the Pioneer and Voyager probes, in order to detect reflected sunlight. A major finding was that the zodiacal light and *gegenschein*, and, therefore, the dust which causes them, are essentially absent beyond Jupiter's orbit.

The inner Solar System is a dusty place. Particles of this dust, liberated fairly recently in cosmological terms from comets, and shepherded largely by the gravitational influence of Jupiter, are the source of meteors seen in our atmosphere. Observations of meteors, indeed, provide an important means of studying this dust. In the next chapter, I take a closer look at the nature of the small particles which rain into the Earth's atmosphere from the zodiacal dust cloud.

CHAPTER 2

Meteoroids and the upper atmosphere

The nature of meteoroids

Most of our knowledge of the dusty material which pervades the inner Solar System comes from indirect methods of observation. The study of meteors, which signal the impact of small particles on the upper atmosphere, for example, allows something of the particles' densities, sizes, and chemical compositions to be determined, without having the actual particles themselves in a laboratory for detailed examination. Even the simple visual observations described in Chapter 4 can add to our knowledge of meteoroids and their behaviour.

Like the light from a star, light from a meteor – provided it is sufficiently bright – can be broken into its component colours using a prism or diffraction grating, and the resulting spectrum photographed. Analysis of meteor spectra reveals emission lines characteristic of several elements including, notably, calcium, magnesium, sodium, silicon, and iron. The levels of emission from each element depend partly on the meteoroid's entry velocity. Fast meteors such as Perseids strongly emit the H and K lines of singly ionized calcium (Ca II), while emissions of neutral iron, magnesium, and sodium (Fe I, Mg I, Na I) are more common in slow meteors, including Geminids.

Attempts have been made to collect the small fragments of material which remain following the disintegration of meteoroids entering the Earth's atmosphere to produce meteors. In a programme under the direction of Donald Brownlee of the University of Washington, Seattle, American U2 reconnaisance aircraft flying at high altitudes have carried dust collectors designed to gather up meteoroid fragments suspended in the atmosphere for closer study. The *Brownlee particles* thus gathered have a fluffy structure consistent with the clumping together of smaller dust grains in the pre-solar nebula.

Eventually, dust from the upper atmosphere settles out to ground level. Some small spherules of material – *micrometeorites* – are widely believed to be remnants of meteoroids that have reached the ground, where they may be collected by special means. Micrometeorites have been collected from the deep ocean floor by drilling, as during the *Challenger* oceanographic expedition in the late nineteenth century. Rainwater is another source, but researchers need to be able to discriminate between micrometeorites and terrestrial contaminants such as metallic spherules thrown into the atmosphere by industrial processes. A more promising source of micrometeorites is pristine ice in remote Arctic and Antarctic regions. Drilled core samples from the ice contain micrometeorites deposited over long time intervals, and are less prone to contamination.

The advent of exploration by spacecraft has led to yet more direct investigation of meteoroidal material. Satellites in Earth orbit have been used to study the arriving meteoroid flux. The American Pegasus satellites, launched during the 1960s, were equipped with meteoroid detector arrays. Spread out like solar panels, these arrays covered large areas, and were used to assess both numbers and sizes of meteoroids.

In addition to providing images of the nucleus of Comet P/Halley, the Giotto mission in 1986 also investigated the size, mass, and chemical composition of meteoroids in the vicinity of the nucleus. Giotto received a considerable battering from such particles as it sped by P/Halley's nucleus. Impacts at a relative velocity of 68 km/s (42 mile/s) caused some damage to Giotto despite the probe being equipped with a "bullet-proof" Kevlar shield. At these relative velocities, there is sufficient energy for even small particles to punch holes through sheets of metal. A collision between Giotto and a 0.1 g particle, probably no larger than a couple of millimetres in diameter, threw the probe into a temporary spin a few seconds before its closest encounter with the comet's nucleus. Further collisions with smaller particles ripped away the protective baffle covering the probe's camera. Giotto's main instruments for studying meteoroids were a pressure detector, by which size and mass could be estimated, and a mass spectrometer. The mass spectrometer was used to examine the chemical composition of solid particles from P/Halley's nucleus impacting on Giotto, and vaporized in its collector. This analysis showed the particles to consist largely of silicates, and also suggested the presence of quantities of organic components in the so-called CHON particles (containing carbon, hydrogen, oxygen, and nitrogen).

All these lines of investigation – from simple naked-eye observations by amateurs to close comet fly-bys using sophisticated spacecraft – come together to produce our current picture of meteoroid structure and composition. Cometary meteoroids are widely thought to consist of small clumps of primordial dust grains, loosely bound together in a lattice structure. In the parent comet, the spaces between the grains are presumably filled, initially, by water-ice and frozen gases which sublime away during the comet's perihelion passages.

The typical meteoroid (which would give rise to a naked-eye meteor of about second magnitude) has a mass of 0.1 g and a size of less than 1 cm. Such bodies therefore have very low densities, accounting for their rapid disintegration in the Earth's atmosphere. Meteoroids disintegrate into the smaller fragments of 10 μm (10^{-2} mm) diameter collected as Brownlee particles. Brownlee particles themselves are made up of yet smaller units, whose size – of the order of 1 μm – is consistent with astronomers' indirect measurements of the size of dust grains in interstellar nebulae. The chemical composition of meteoroids (silicates, with traces of the heavier elements) is also consistent with their origin as grain-clumps, bound together by ices, and preserved from the beginnings of the Solar System in comets until their release, aeons later, into meteor streams.

**The upper
atmosphere**

It is difficult to say exactly where the Earth's atmosphere peters out into the near-vacuum of interplanetary space. Observations from spacecraft show the planet to be surrounded, out to a distance of at least four Earth radii (over 25,000 km, 16,000 miles), by an extremely tenuous cloud of hydrogen called the *geocorona*. For most practical purposes, however, it is more appropriate to consider the thin traces of atmosphere 100–300 km (60–200 miles) above the Earth's surface as representing the uppermost fringes. Many artificial satellites orbit at around these altitudes, and are directly influenced by the atmosphere's behaviour. Most meteors occur at an altitude of 80–110 km (50–70 miles), where the atmospheric density becomes more significant.

The precise altitude of this *meteor layer* depends largely on the sunspot cycle. Sunspots are regions of intense magnetic activity on the visible surface of the Sun, appearing dark because they are slightly lower in temperature than the surrounding photosphere. Sunspot numbers rise and fall over time, reaching a peak roughly every 11 years (Figure 2.1). Sometimes pent-up energy is violently released from sunspot regions in *solar flares*. Such flares emit considerable amounts of short-wavelength ultraviolet and X-ray radiation, which heat the

Figure 2.1 *Solar activity, as manifested by sunspot numbers, shows a cycle of rise and fall, with successive sunspot maxima at intervals of about 11 years. Plotted here are monthly values for the mean daily frequency of sunspot groups, observed between 1978 and 1991. Maxima occurred in 1979 and 1989–90, with a decline in the more recent cycle, number 22, setting in during the second half of 1991. The cycles are numbered from when systematic records first began to be kept.*

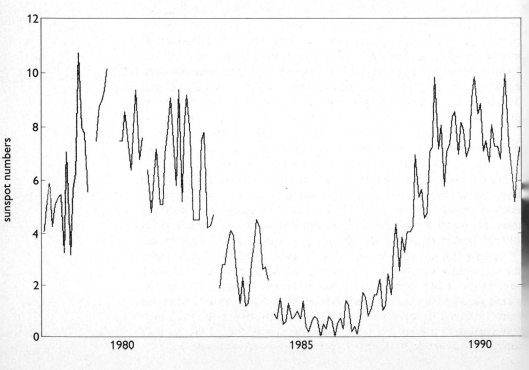

Earth's upper atmosphere. Around the time of sunspot maximum, flares may occur as frequently as 25 times a day.

Sunspot maximum normally lasts about 18 months, but can be longer: for example, sunspot cycle 22 had a long maximum, peaking between 1989 and 1991. The effect of such activity is for the Earth's upper atmosphere to become more extended (in other words, like any other gas subjected to heating, it expands). As a result the atmospheric density rises at greater altitudes. Conversely, as sunspot activity dwindles towards minimum, the upper atmosphere contracts. A consequence of these changes in the extent of the upper atmosphere is that the meteor layer appears to move up and down in response to sunspot activity. The atmospheric density becomes sufficiently high for meteoroids to be destroyed by collisional heating at greater altitudes at sunspot maximum than at sunspot minimum.

This effect also has important implications for artificial satellites in low Earth orbit. Such satellites experience a certain degree of drag as they pass through the upper atmosphere's outer fringes. Obviously, the more extended the atmosphere, the greater the drag. Atmospheric drag robs a satellite of orbital momentum, thereby causing its orbit to drop progressively lower. Unless boosted back to a higher altitude, a satellite's orbit will eventually "decay," to be destroyed – like an incoming meteoroid – by heating during re-entry. Extremely high levels of sunspot activity in 1989 brought about the premature deaths of several satellites including, ironically, the Solar Maximum Mission satellite, measurements from which, during its nine years of operation, provided scientists with many valuable insights into the Sun's activity.

The principal constituents of the Earth's atmosphere are nitrogen and oxygen. At sea level, these are mixed in the ratio 78 percent nitrogen to 21 percent oxygen, with other gases such as carbon dioxide in trace quantities. The bulk of the atmosphere is held close to the Earth's surface in the troposphere, the layer in which "weather" clouds are found, which extends up to a maximum altitude of 15 km (9 miles). The atmospheric density falls off with increasing altitude in the troposphere, and more markedly so in the overlying stratosphere.

In the stratosphere, at altitudes of around 50 km (30 miles), the action of solar ultraviolet radiation begins to dissociate oxygen molecules (O_2) into their two constituent atoms (O + O). Free oxygen atoms can combine with molecular oxygen to produce ozone (O_3; $O_2 + O \rightarrow O_3$), and the ozone layer thus formed acts as an important screen against the solar ultraviolet radiation which creates it. The process of photodissociation becomes increasingly significant higher in the atmosphere: above about 130 km (80 miles), almost all atmospheric oxygen is in the form of single atoms.

Around 100 km (60 miles) up, at the altitude of the meteor layer, the atmosphere consists of 76.5 percent molecular nitrogen (N_2), 20.5 percent molecular oxygen, and 3 percent atomic oxygen. The atmospheric density here is extremely low, about a millionth of that at sea level – comparable to the gas density in a domestic light bulb. Even such low densities, however, are sufficient to cause the rapid destruction of meteoroids impacting at high velocities on the upper atmosphere.

The meteor process

The events which give rise to a visible meteor at heights of around 100 km (60 miles) above the Earth's surface take place over a very short timescale. Typically, a meteor lasts only a couple of tenths of a second, particularly if the meteoroid impacts on the atmosphere at a high velocity. More rarely, slow meteors may last as long as one or two seconds.

An incoming meteoroid collides with increasing numbers of atmospheric atoms and molecules at altitudes between 115 and 85 km (70 and 50 miles). Atmospheric particles become attached (*adsorb*) to the meteoroid's surface in collisions which impart energy as heat to the meteoroid via friction. Atoms are displaced from the meteoroid's surface by this process; each adsorbed atmospheric particle ejects about a hundred meteoroid atoms. The meteoroid surface is rapidly eroded by this process of *ablation*. As the surface is vaporized, the meteoroid's lattice of small grains breaks up rapidly into what is, effectively, a small cloud of material undergoing ablation. Atoms ejected during ablation of a meteoroid collide with further atmospheric particles, producing excitation and ionization.

To understand the processes of excitation and ionization, it is necessary to take a brief look at atomic structure. Each atom consists of a positively charged *nucleus*, surrounded by a cloud of negatively charged *electrons*, and has an overall electrical charge of zero. The number of electrons and the composition of the nucleus (containing protons and neutrons) depend on the chemical species.

In many popular-science texts the atom is depicted as a miniature solar system, with electrons orbiting the nucleus just as planets orbit the Sun. For simple purposes this model is adequate, provided its limitations are realized. An important difference is that those electrons farthest from the nucleus orbit the fastest, the complete opposite of the planetary situation. Electron orbits can lie only in certain permitted shells, whose distances from the nucleus are determined by the rules of quantum mechanics. These laws also dictate that the lower-energy inner orbitals must be filled before electrons can occupy the more energetic outer orbitals. A precise amount of energy – a *quantum* – must be imparted to the atom in order to send an electron from a lower to a higher orbital. Such energy may come, for example, from collisions with other atoms in a heated gas, or from energetic radiations such as ultraviolet or X-rays from the Sun.

Excitation will raise electrons to higher energy levels, a situation which is stable for only a short time. The excess energy imparted to the atom is often quickly lost, being re-radiated in a small fraction of a second as a quantum packet of light. It is this process which causes the short-lived streak of light seen as a meteor. Atoms ejected from the surface of the ablating meteoroid produce a 20–30 km (12–20 mile) long column of excitation a few metres wide in the upper atmosphere. Excitation both of meteoroid particles and atmospheric particles (oxygen and nitrogen) contributes to the visible phenomenon of a meteor.

More energetic collisions lead to *ionization*, where electrons are not just kicked into higher orbitals, but are stripped away from the nucleus altogether. The process of *recombination*, in which such a positively ionized nucleus captures free electrons to become electrically neutral again, also leads to the emission of light, seen as *persistent trains*, or – if of extremely short duration – *wakes*, behind the brightest and fastest meteors.

The important factors which contribute to a meteor's brightness are the size of the impacting meteoroid, and the velocity at which it arrives in the upper atmosphere. Most meteoroids, arriving with a *geocentric velocity* between 11 and 72 km/s (7 and 45 mile/s), undergo negligible deceleration by the atmosphere before being completely destroyed by ablation. The energy (E) of such an impact is given by the familiar formula $E = \frac{1}{2}mv^2$, where m is the mass of the meteoroid and v is its velocity. Clearly, more massive and/or higher-velocity meteoroids will produce higher-energy collisions, and, in general, the higher the energy, the brighter the meteor. Persistent train phenomena, resulting from ionization by energetic collisions, are most often seen for meteoroids – such as those from the Perseid, Leonid, or, especially, Orionid streams – which arrive at very high velocities.

Gerald S. Hawkins in his classic 1964 work *Meteors, Comets and Meteorites* provides a detailed mathematical treatment of the physical processes giving rise to meteors. He concludes that the vast majority of the collisional energy in these events goes to heating the incoming meteoroid. For slow meteors, light emission accounts for only 0.01 percent of the energy budget, and ionization a further order of magnitude less. Fast meteors emit 0.1 percent of the collisional energy as light, while 0.005 percent of their energy goes into ionization.

Diurnal and seasonal effects on meteor rates: the sporadic background

Most basic textbooks on general astronomy will assure the reader that "on any clear night, about six or seven meteors can be seen every hour." Anyone who has tried meteor work during the quieter months of the spring will recognize this statement as a myth: presumably the authors of such books are not regular meteor observers! I have known intervals of up to an hour and a half to go by, meteor free, during watches in the early spring, and this is clearly not the best time of year for a beginner to take up meteor observing.

There are a number of reasons why meteor activity should vary from one time of year to another, and even over the course of a single night. The most obvious is the existence of distinct showers – rates will be higher while the Earth is crossing a rich meteor stream such as the Perseids. Away from shower activity, however, the flux of random background sporadic meteors still shows hourly and seasonal variations. Why should this be?

Sporadic meteor rates are lowest in the early evening, rising gradually through the night to higher levels just before dawn (Figure 2.2). This *diurnal variation* is observed every night, and is accounted for by a combination of the Earth's rotation and its orbital motion. In the early evening the observer is on the trailing hemisphere of the Earth relative to its direction of orbital motion around the Sun. Most sporadic

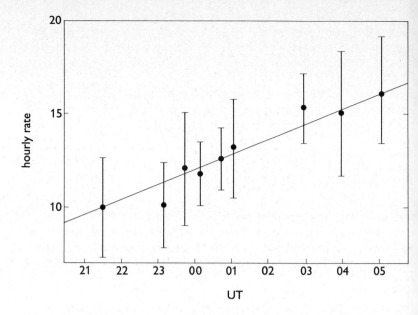

Figure 2.2 *Observed sporadic rates show a gradual increase as the night wears on, reaching their highest around dawn. Averaged figures for this diurnal variation are used to calibrate shower rate analyses as described in Chapter 4. The observations shown here relate to the peak period of December 11–15 during the 1990 Geminid shower. The vertical bars in this and other diagrams represent estimates of error in the results.*

meteors seen at this time of night are the result of meteoroids "catching up" with the Earth from behind; the solid bulk of the Earth lies between the observer and the head-on collisions on the leading edge. By dawn, the planet's rotation has carried the observer round to the leading edge, where sporadic meteoroids are swept up as the Earth ploughs through the zodiacal dust cloud (Figure 2.3). This is rather like a vehicle travelling through a gentle snowfall – the front of the vehicle collects more snowflakes than the sheltered rear. Sporadic meteoroids catching up with the Earth during early evening impact on the atmosphere at relatively low velocities, and are consequently not only less frequent, but also fainter on average, than those seen just before dawn.

Shower meteors, as we saw in Chapter 1, appear to emanate from a relatively small area of sky, the radiant. Sporadic meteors, on the other hand, are regarded as being more or less random in their direction of origin. The Earth's orbital motion carries it through the dusty background of the inner Solar System, sweeping up meteoroidal material ahead of it as it goes. Many sporadic meteors come, therefore, more or less from the direction in which the Earth is heading.

The point in the sky towards which the Earth's orbital motion is apparently carrying it is called the *apex of the Earth's way*, and lies 90° westward along the ecliptic from the Sun. Around the autumnal equinox, the apex lies at its highest point along the ecliptic, on the Gemini/Taurus border, where the summer solstice Sun lay around June 21. Conversely,

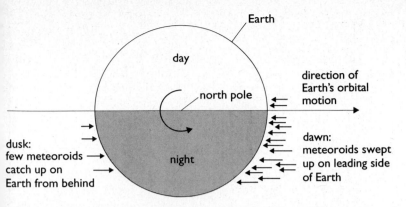

Figure 2.3 *The diurnal variation shown in Figure 2.2 results from the differing position of the observer with respect to the Earth's leading edge, where most sporadic meteors are swept up. In early evening the observer is on the trailing edge, and is carried towards the leading edge – towards dawn – by the Earth's rotation, as shown in this "top" view.*

during the spring the apex is at its lowest declination, in Sagittarius, at the position of the winter solstice Sun.

More of the sporadic meteors resulting from debris swept up from the space ahead of the Earth will be observed when the apex of the Earth's way is high in the sky than when it is low. Consequently, observed sporadic rates in the spring are about a third to a half of those recorded in the autumn (Figure 2.4). A typical hour's observing in March might yield three or four sporadics at best, compared with eight to ten in September.

Figure 2.4 *Sporadic meteor rates show an annual variation, being on average perhaps twice as high in the autumn months as during the spring.*

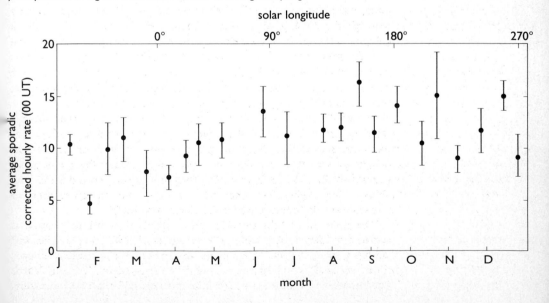

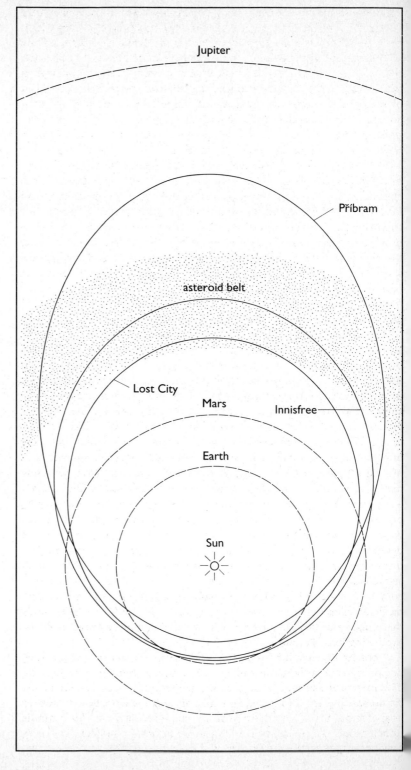

Figure 2.5 *The meteorites which were recovered at Lost City, Přibram, and Innisfree were detected by fireball patrol networks as extremely bright meteors during descent. Multiple-station photography yielded precise measurements of position, which were used to determine the original Solar System orbits for these bodies. All three can be traced back to the main asteroid belt, which is believed to be the principal source of meteorites landing on Earth. (See p. 60.)*

The "average" meteor seen by the visual observer is usually of second or third magnitude, about the same brightness as the stars which make up the familiar asterism of the Plough, or Big Dipper. More rarely, an observer who puts in long hours of meteor watching will be rewarded by the sight of very much brighter events, perhaps even comparable in brightness to the Moon. Many of the meteors noticed by members of the general public are such events, sometimes associated with the extremely rare arrival of bodies sufficiently large to survive atmospheric passage and reach the ground as meteorites. Such events are, rightly, newsworthy, and may cause great excitement if widely seen.

Fireballs

Meteor astronomers use the term *fireball* to describe those objects brighter than the planet Venus (magnitude −5 or brighter). Fireballs are comparatively rare, and an observer may see only one or two such events in a year's meteor work. The major annual showers – in particular, the Perseids, Geminids, and Quadrantids – produce quite a few very bright events, and the chances of observing a fireball are greater around the times of their activity peaks. Fireballs are also found among the sporadic background. Long-term studies suggest that certain times of year are more favourable than others for sporadic fireballs. The period from February to April has a reputation for being the "fireball season" for northern hemisphere observers.

The fireballs seen during major showers are produced by the atmospheric impact of larger, and therefore more massive, "dust-ball" meteoroids within a meteor stream. Like their smaller counterparts, these bodies very rapidly disintegrate completely. Some of the randomly occurring fireballs seen at other times, however, are produced by more substantial particles. Rarely, such particles may survive the ablation process and fall to Earth as meteorites. The principal source of such objects is widely accepted to be the main asteroid belt between the orbits of Mars and Jupiter, where occasional collisions result in fragmentation of asteroids, and can eject small shards of material – perhaps up to a few metres in diameter – into orbits which cross that of the Earth (Figure 2.5).

By strict convention, these bodies are still referred to as meteoroids, but they have a much different composition to those which are found within cometary meteor streams. In nature, asteroidal fragments are denser than cometary meteoroids, making them less prone to complete disintegration should they enter the Earth's atmosphere. These small asteroidal fragments are still subject to the same processes which destroy cometary meteoroids in the atmosphere: adsorption of atmospheric particles causes ablation and evaporation of the outer surface. Since these bodies are larger they can withstand ablation for longer periods, producing long-duration, very bright meteors.

A potentially meteorite-dropping fireball may last for several seconds, beginning as a fairly modest meteor, but rapidly gaining in brightness. The meteoroid surface, as it becomes heated, produces a molten shell on the leading edge from which fragments become detached, to appear as sparks or "blobs" in the trail behind the meteor's head. Fluctuations in brightness may occur as material is shed from the meteoroid. Flares may even light up the ground, casting shadows.

Some asteroidal fragments have a significant cross-section, and – unlike their smaller counterparts – undergo marked deceleration by air resistance in the high atmosphere. Sudden deceleration may generate sufficient internal stress to cause the incoming body to break up abruptly, giving rise to a major flare in brightness. Such flares are bad news for anyone hoping to recover an intact meteorite on the ground!

Fireballs associated with the arrival of larger, more solid objects can penetrate to quite low atmospheric levels, remaining luminous to within 20 km (12 miles) of the ground. Explosions at these altitudes can produce sonic booms, rumbles, and other sound effects detectable at ground level. Atmospheric pressure waves from some low-penetrating fireballs have been detected by seismometers. Fragmentation of large meteoroids in the atmosphere can leave behind *trails* of debris which remain visible for several minutes; in extreme cases these trails may persist for several hours, becoming distorted by stratospheric winds.

Reports of possible meteorite-dropping fireballs are welcomed with great interest by researchers. Although there are large collections of meteorites in museums around the world, each new arrival has the potential of representing a sample from a previously unstudied class. For many known meteorite classes – particularly the carbonaceous chondrites, rich in organic compounds – early collection of the fallen object is desirable, before terrestrial contamination is introduced or the meteorite is exposed to weathering. Photographic patrol networks designed to record bright fireballs with the ultimate aim of rapidly collecting any resulting fallen meteorites have been operated, with some success, in a number of countries (see Chapter 3).

Visual reports, even from untrained members of the public, can be of value in reconstructing a fireball's atmospheric trajectory, and determining the likely fall zone for any resulting meteorites. More often than not, sadly, those occasional fireballs which are widely seen result in complete destruction of the incoming body, and nothing can be recovered. The rare occasions on which a meteorite can be found make the pursuit of all possible eyewitness accounts of major fireballs worthwhile.

Regular visual observers should always be aware of the possibility of a fireball occurring during their watches, and be ready to take note of the appropriate details. Systematic photographic patrol work (sadly neglected since its heyday in the 1960s) can also have its rewards. Both these observational aspects will be discussed later.

Meteorites

The study of meteorites is a scientific discipline almost to itself. The majority of meteor events seen by amateur observers result from small particles of cometary debris which, as stated earlier, have no chance of reaching the ground intact. Most of the larger bodies which do survive their plunge through the atmosphere to reach the surface are believed to be of asteroidal origin, on the basis of computations of original orbits for incoming meteorites photographed from different locations during their atmospheric passage as fireballs. Meteorites therefore give us the opportunity to examine primitive Solar System material at first hand, and while in many ways they are quite distinct from the debris which produces ordinary meteors, it is worth taking a small diversion to consider these bodies.

Meteorites are collected by two means, and may be described either as falls or finds. A *fall* usually results from a fireball event which has been well observed, resulting in an accurate determination of the likely drop-zone for any remaining debris. Meteorites, usually entering the atmosphere at glancing angles, often scatter debris over an elliptical area, referred to as a *strewnfield*. Subsequent searches of the strewnfield can lead to the collection of the fallen object, as, for example, with the Barwell Meteorite, which fell in Leicestershire, England, on 1965 December 24, and the Allende Meteorite, which fell in Mexico on 1969 February 8.

More rarely, eyewitnesses may be near the actual area where the meteorite landed. The most recent documented fall in the UK – the first since 1965 – was in 1991 May, when the Glatton Meteorite fell near Peterborough, in the garden of Arthur Pettifor. It was altogether rather fortuitous that this fall was noticed at all: at the time Mr Pettifor was tending his onions, and he noticed the disturbance caused by this small object, measuring only 100×60 mm (4×2 inches) crashing into a nearby hedge. Conditions were cloudy over most of the country on the day of the fall, and no one saw the (daytime) fireball which should have marked the meteorite's arrival.

There must be a fair number of such falls which go unnoticed in the course of each year. Only when they have other effects, such as loud sonic booms or (fortunately extremely rarely) structural damage, might unobserved falls draw attention to themselves. One well-documented instance was the fall of a stone meteorite – following a fireball sighting – on the evening of 1992 October 9, which smashed through the rear wing of a parked Chevrolet in Peekskill, New York. Statistically, falls are most frequent during the afternoon.

More commonly, the meteorites which grace the collections of museums around the world are *finds*, made long after the object has fallen. Again, a fair degree of chance is involved in such meteorite discoveries. Finds may include "stones" which have been removed from fields by farmers, or meteorites which have lain in remote desert areas, undisturbed since they fell, to be discovered much later. A number of meteorites were found by explorers in the eighteenth and nineteenth centuries. Many of these objects were known to the original inhabitants of the region, who sometimes looked upon the fallen meteorites as religious artifacts. For example, the explorer Robert Peary returned to New York in 1897 with a group of three meteorites discovered at Cape York, Greenland. Known to the local people as Ahnighito (the Tent), the Woman, and the Dog, they were parted with only reluctantly. Ahnighito, a 31-tonne mass, had been used as a source of iron before the days of trading with whalers.

Among others discovered during the age of world exploration is the world's largest known intact meteorite, which lies half-buried where it fell at Hoba West in southern Africa. The Hoba West Meteorite is a mass of iron, weighing at least 60 tonnes. It was once believed that a great mass of meteoritic iron might lie under the famous Meteor Crater in Arizona. Extensive drilling and other geological investigations have revealed no trace of this, however, implying that the incoming meteorite responsible for producing the crater was largely vaporized on impact. Fragments of meteoritic material have been found scattered in the desert around the crater, lending support to this view.

More recently, the Antarctic ice has proved a rich collecting-ground for meteorites. As the ice sheet flows outwards from the interior of the continent towards the Southern Ocean, it carries with it meteorites which have fallen on it over the several millions of years of its existence. In certain areas, regions of reduced ice mobility (rather like eddies in a river) develop, and meteorites carried along in the ice can become trapped in these areas. Wind and driving snow gradually scour away the ice surface, eventually exposing the meteorites. Since the late 1970s, researchers from the United States and Japan have explored the ice flows, searching for meteorites to study.

In all, something like 3000 meteorite falls are now represented in collections worldwide, complemented by a further 7000 objects recovered from the Antarctic ice. They are usually named for the location at which they were found.

Meteorites are often identifiable in having a dark fusion crust, produced during ablation in the atmosphere. This crust often bears markings which indicate that the molten exterior of the meteorite flowed backwards during the fall. Contrary to popular belief, a meteorite is often cold to the touch if picked up soon after falling: only its outermost layer becomes heated by its atmospheric passage, the interior remaining at the low temperatures of interplanetary space, and conduction ensures that the outside cools rapidly after the fall. Meteorites can sometimes be distinguished from native rocks in the fall area by their fusion crust (though investigators have to be wary of industrial slag), and by their greater density. A meteorite will often be heavier than a terrestrial rock of the same size.

There are two main classes of meteorite: the *irons* and the *stones*. Iron meteorites consist principally of a nickel–iron alloy, while stony meteorites have a composition dominated by minerals such as olivine. Intermediate between the two classes are the *stony–irons*, which, as their name suggests, contain significant quantities of both nickel–iron and minerals. Among finds, the irons tend to be better represented – perhaps because they are more resistant to weathering.

Mineralogical investigations of meteorites reveal much about their nature and possible origin. Iron meteorites, when sectioned, polished, and etched, show characteristic angular patterns resulting from crystallization of the nickel–iron alloy at relatively low pressure, as might be expected within a small (asteroid-sized) body. In particular, the *Widmanstätten pattern* of octahedrites is found only in meteorites. Other iron meteorites show a hexahedrite pattern.

Cross-sections through stony meteorites often reveal a *chondrite* structure: they are made up of a matrix containing small inclusions – chondrules – bound together. Stony meteorites can be subclassified by their iron content; all meteorites contain at least some quantity of iron. Chondrites are believed to be meteorites which condensed from the pre-solar nebula (see Chapter 1), and have not undergone much subsequent modification (by incorporation into a larger body, for example), though they may have lost any volatile components originally present.

By contrast, a second class of stony meteorite, the *achondrites*, are believed by many scientists to have undergone some degree of secondary processing. Examples include the uncommon subclasses of meteorite known as eucrites, shergottites, nakhlites, and chassignites. These show evidence of having once been parts of larger bodies in which they had become molten, perhaps during the process of *differentiation*.

Differentiation occurred in the Earth fairly early in its history, with the result that heavier materials (iron and nickel, for example) sank towards the planet's core, while lighter materials such as the basaltic minerals "floated" to form the continental crust. It is quite likely that similar processes acted on the other terrestrial planets, and among the larger planetesimals in the asteroid belt. Iron meteorites may represent fragments of core material from partly differentiated planetesimals which were destroyed by cataclysmic collisions in the asteroid belt. Spectroscopic studies have led to the suggestion that eucrites may represent fragments of crustal material from a body similar to the asteroid Vesta, while shergottites, nakhlites, and chassignites, known collectively as *SNC meteorites*, may be pieces of Martian crust. SNC meteorites are believed to have their origin in large meteorites impacting on Mars and ejecting material from the planet's surface into Earth-crossing orbits. Similar impacts on the Moon were presumably the source of the lunar material found among the meteorites from the Antarctic.

A final, important class of meteorite is the *carbonaceous chondrites*. These extremely friable meteorites contain, as their name suggests, substantial amounts of carbon, including organic compounds. Among these compounds are amino acids, one of the two main classes of biological precursor molecule. Some astronomers have suggested that such meteorites may have had an important role to play in the early development of life on Earth, but it is only fair to mention that most biologists remain highly sceptical. None the less, the presence of these materials in carbonaceous chondrites points to the existence of several interesting chemical processes in the pre-solar nebula, and the study of these rare meteorites is important to our developing understanding of conditions in the very early Solar System.

Quite distinct from meteorites are the glassy *tektites*, found in strewn-fields covering great areas. Examples include australites from Australia, and moldavites from the eastern European plain. Tektites are thought not to have originated from beyond the atmosphere, but to be secondary products of giant meteorite impacts. As with Meteor Crater in Arizona, the incoming bodies were largely vaporized on impact. These events were sufficiently energetic to melt areas of the Earth's crust, ejecting the molten rock away from the impact site. Cooling and solidifying in flight, the ejected material came down far from its source as tektites.

Study of the Earth's surface from satellites has revealed a number of possible impact sites from which the known tektite fields might have originated. These large circular features are known as *astroblemes* – literally, "star wounds" – and bear witness to impacts by large bodies (possibly including Apollo asteroids, as discussed in Chapter 1) in the distant past. The extent of erosion of these features points to their great age.

Meteorites play a fundamental role in our understanding of how the Solar System came together, and it is for this reason that amateur fireball observations, with the possibility of rapid recovery of any resulting solid material on the ground, are so useful.

Man-made fireballs

Some fireballs result from the re-entry of artificial satellites, or parts of rockets used to launch them. The orbital environment about 300 km (200 miles) up is becoming increasingly cluttered with space-age debris, some of which eventually decays from orbit to produce spectacular fireballs. From his home in Hawaii, William Albrecht is well placed to observe the re-entry of the spent fuel tank following a launch of the Shace Shuttle, and has photographed many of the spectacular displays put on by the disintegrating tanks. A number of Soviet Cosmos satellites contained magnesium in their welding, resulting in particularly bright re-entry events. Re-entries are usually very slow-moving compared with high-velocity natural events, and reports often mention gradual fragmentation and long durations.

Fragments do occasionally make it back to Earth, as in the celebrated cases of the Skylab and Salyut 7 space stations, large pieces of which crashed into the Australian desert and the Andes, respectively. Smaller pieces are also recovered from time to time, such as the fragment of a Cosmos rocket which crash-landed on a golf course at Eastbourne in southern England in 1979 January.

CHAPTER 3

A brief history of meteor studies

Not surprisingly, meteors have been known for millennia. Our sky-watching ancestors from Babylonian times (4000 BC) and earlier, who gave some of the stars and constellations the names familiar to today's meteor observers, can hardly have failed to notice occasional "shooting stars" during their nightly vigils. Systematic records are, however, rare, and it is not until around 2000 BC in China and Korea that the first indications of meteors being recorded as astronomical phenomena and astrological portents are found.

Successive Chinese emperors saw the advantage of having an Astronomical Bureau, charged with recording the various heavenly signs. In particular, following the installation of a new emperor it was advisable to have the celestial signs point to a long and fruitful reign. In this connection there are surviving written records of astronomical phenomena such as novae and supernovae, aurorae, and noteworthy showers of meteors.

Such records have been used by Jack Eddy, a solar astronomer at Boulder, Colorado, to trace past interludes during which the Sun may have been more, or less, active than at present. The incidence of aurorae at lower latitudes can be correlated reasonably well with increased sunspot activity. Chinese records of sunspots (observed with the naked eye through thick haze near sunrise or sunset) and aurorae point to higher-than-present solar activity in the period around 200 BC, and again during the twelfth and thirteenth centuries AD, evidence corroborated by European records for the latter period.

Likewise, meteor activity can sometimes be followed into past ages. Records of nights on which, for example, "More than 100 meteors flew thither in the morning" reveal that the Perseids, a current favourite shower for amateur observers, were active at least as long ago as AD 36. Other showers are similarly recorded. Examples include the Lyrids in 687 BC: "Stars fell like a shower"; the Eta Aquarids in AD 466: "Countless large and small meteors flew west"; the Orionids in AD 585: "Hundreds of meteors scattered in all directions"; and the Leonids in AD 1002: "Scores of small stars fell."

Changes in the pattern of activity can be seen. For example, during the Middle Ages the Taurids were possibly as active as today's Perseid shower, though now they are rather weak, and tend to be observed only by the most dedicated, patient amateur meteor watchers. The

Quadrantids, which currently rank as one of the three most active showers of the year, went missing for a time in the Middle Ages, a phenomenon now accounted for by the oscillation up and down of the stream's orbit relative to the ecliptic plane as a result of gravitational perturbations by the giant planet Jupiter. Computer modelling has shown that in a few hundred years the Quadrantids will again miss the Earth, and the shower will be absent from future generations' skies.

Useful as they are in tracing past meteor and other activity, the Chinese historical records must be regarded with some caution. At times, it was politically expedient for celestial harbingers to go unrecorded, and there are a number of events – well documented elsewhere – which did not find their way into the oriental annals.

European records can also provide some insights into meteor activity in the distant past. In the *Anglo-Saxon Chronicle*, a collection of monastic writings on events in England dating back to Roman times, there are a couple of references to meteor activity. One example, from 1095, relates how

> at Easter, on the mass night of St Ambrose [April 4] there were seen, nearly over all this land, and nearly all the night, very many stars as it were to fall from heaven, not by one or two, but so thickly that no man could count them.

A later account, by Matthew Paris in his *Historia Anglorum*, describes Perseid activity in 1243:

> This year, July 26, in a perfect clear night, behold, bright stars were seen to fall, shooting here and there, and had they been true stars falling none would be left in the sky.

Later records: European fireballs

Local chronicles from post-Renaissance Europe contain records of day-to-day life, and occasionally note unusual weather conditions, or even the rare occurrence of a bright fireball (often referred to as *draco volans* – a flying dragon) or great displays of meteors. This example is from an early-eighteenth-century parish record from the village of Staunton in Worcestershire:

> Upon Thursday the 19th of March about eight o'clock at night, the Moon shining very bright when was a sudden lightning in the air attended with a glowing Heat which continued for the space of half a minute so very light that the Moon and stars were not visible. It ran in streams from the North East to the South West where it remained only like a small white cloud for about half an hour after. It was a surprise to many spectators and reckoned a very extraordinary meteor. About 2 minutes after was heard a great howling noise in the air and about 5 minutes after such a crack was heard as could not be made by the largest cannon.

A Northumbrian record from 1750 July 22 reads:

> Between eight and nine at night a strange phenomenon was seen in the air, it appeared first at a considerable distance in the north, passed this place with great velocity to the south, and seemed to be a body of fire

about twenty inches [50 cm] round, and in its motion had a luminous tail about two yards [2 m] long.

A few years later, on 1758 November 26, another chronicler in Newcastle tells of how

> This night a surprising large meteor was seen at Newcastle, about 9 o'clock, which passed a little westward of the town, directly north, and illuminated the atmosphere to that degree, for a minute, that, though it was dark before, a pin might have been picked up in the streets. Its velocity was inconceivably great, and it seemed near the size of a man's head. It had a tail of between two and three yards [2–3 m] long, and as it passed, some said they saw sparks of fire fall from it.

The frequency of such accounts has been used by some investigators in attempts to reconstruct the past activity of meteor showers. Although many fireballs result from the impact on the atmosphere of isolated fragments of rocky debris, some are associated with the major annual showers. From the past frequency of fireballs, it may be possible to infer the past activity of more "normal" meteors within a shower. To this end, several catalogues of historical fireballs have been prepared. Much library research is necessary, however, to unearth these occasional sightings, and this is an area where progress can be extremely slow.

A particularly interesting fireball event occurred on 1783 August 18, in the early evening. This extremely bright, fragmenting object passed southwards across the British Isles, and on towards France, being seen by several witnesses. The physician Charles Blagden (1748–1820) carried out a detailed investigation of its behaviour, and a number of reports subsequently appeared in the journal *Philosophical Transactions of the Royal Society*. The event was recorded for posterity in an etching by Henry Robinson of Newark, and in a painting by Paul and Thomas Sandby, who saw the fireball from Windsor Castle. A watercolour of the fireball by the Sandby brothers realized £26,000 when sold at auction in 1989.

Scientific studies of meteors only really got under way in the late eighteenth century. Before this time, few had made much effort to try to understand the phenomenon. In the fourth century BC the Greek philosopher Aristotle had mentioned meteors in his work *Meteorologica*, rightly treating them, along with aurorae and clouds, as atmospheric phenomena. (Slightly less accurate was his inclusion of comets in the same category.) Meteors, in Aristotle's view, were the result of "hot exudates" being driven off the Earth by the Sun during the day, which, on rising to the lowest of the celestial spheres, became ignited by friction. Like many of Aristotle's views this theory long remained unopposed, partly on religious grounds. His model of a perfect Universe was compatible with the philosophy of the powerful pre-Renaissance Catholic Church, and could not be lightly challenged.

Another popular view from ancient times was that meteors were residual rays of sunlight, appearing after dark. Following the great fireball

The beginnings of scientific meteor study

of 1783, Blagden speculated that meteors might be an electrical phenomenon in the high atmosphere. Indeed, even at this time, few people were convinced that meteors could originate from beyond the atmosphere.

In the late eighteenth century the question of what meteors *really* are, and where they occur, was investigated in detail by two German students, Heinrich William Brandes (1777–1834) and Johann Friedrich Benzenberg (1777–1846), who laid some of the foundations for subsequent studies. At the University of Göttingen in 1798, they decided to attempt simultaneous observations of meteors from sites some 10–15 km (6–9 miles) apart. Using the effect of parallax, by which the same meteor would appear at a slightly different position against the background stars from the two sites (Figure 3.1), they were able to conclude, correctly, that meteors occur at altitudes in excess of 35 km (20 miles). Their results, obtained using only the naked eye, were compromised somewhat by the short baseline, so much so that a handful of the meteors were actually computed to be travelling upwards from the Earth; the issue of meteoric origin remained to be settled. Brandes and Benzenberg published their results in 1800, in a paper entitled "Versuche die Entfernung, die Geschwindigkeit und die Bahnen der Sternschnuppen zu bestimmen" ("Attempts to Determine the Distance, the Speed, and the Paths of Shooting Stars").

The late eighteenth century was also a time when exploration of the Earth was well under way. During a five-year expedition to South America, the German naturalist Alexander von Humboldt and the

Figure 3.1 *The principle of triangulation, as used by Brandes and Benzenberg to determine meteor heights. Observers at A and B, two locations separated by a reasonable distance (preferably over 30 km, 20 miles), will see a meteor against a different part of the sky. Trigonometry can be applied to determine the start and end altitudes, and thus an actual path, for each meteor observed in this way.*

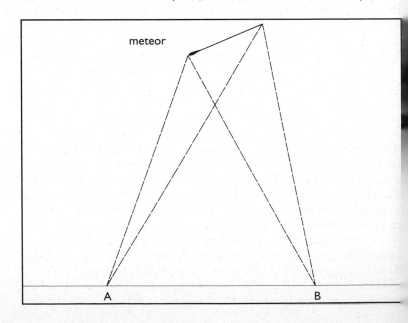

French botanist Aimé Bonpland were fortunate to witness the spectacular Leonid meteor storm of 1799 November 11. Rising early to take the pre-dawn air, the explorers found the clear sky laced with thousands of Leonid meteors. This display was also seen, just before dawn, by observers in the British Isles. Research by the Scottish astronomical historian David Gavine has uncovered British accounts of the Leonid storm, from a number of locations, including Banffshire in Scotland, Hartlepool, Enfield, and the New Forest in Hampshire, in several journals of the day such as *The Gentleman's Magazine*.

Humboldt and Bonpland later discussed their chance observation with local people, and learned that similar displays had been seen, on a fairly regular basis, in past years. Reporting the observations, Humboldt drew scientists' attention to this roughly 30-year periodicity, and also pointed out that the meteors appeared to emanate from a single part of the sky, which we now call the radiant. Future displays of intense Leonid activity could now be anticipated, and observers waited with interest for the expected return of the 1830s.

They were not to be disappointed. In 1833, there was an immense Leonid storm on November 12 (Figure 3.2). The display, with estimated rates of 200,000 meteors in a 6–7 hour interval, was seen from several locations between the West Indies and Canada. In America, the display caused considerable alarm among many witnesses who believed that the biblical Day of Judgement was upon them! More rational observers, among them the Yale professor Denison Olmsted, again noted the radiant effect, and made attempts to measure the position of the point in the sky from which the meteors appeared to come. Olmsted could see that the radiant moved across the sky at the same rate at which the stars were carried round by the Earth's rotation. Such observations during the 1833 Leonid storm at last provided a convincing demonstration of the extraterrestrial origin of meteors.

The comet connection

While the periodic Leonid storms attracted great interest, there was still little systematic meteor work in progress. Much of what was known was based on hearsay and chance observations. The regular appearance of the "August meteors" – the Perseids – was, however, well known by the early nineteenth century, an observation which has been attributed to Brandes. In 1839, the Belgian astronomer and statistician Adolphe Quetelet (1796–1874) published the first extensive catalogue of meteor showers, based on both medieval and contemporary observations; several other workers later produced similar catalogues. Also, by 1857 the diurnal variation in meteor rates (see Chapter 2) had become recognized.

In 1848, a commission of the British Association published a report of chance meteor sightings collected over the previous nine years. At this time there was still considerable debate about where meteors came from: did they originate on the Moon (as proposed by Pierre Simon de Laplace), or in interplanetary space, for example? In 1861, the US astronomer Daniel Kirkwood (1814–95) – best known for his studies of resonance orbits, now known as Kirkwood gaps, in the

Figure 3.2 *The great Leonid storm of 1833, portrayed in a woodcut prepared some years later. The artist was clearly impressed by the huge numbers of meteors falling from the sky during the display.*

asteroid belt – was the first to propose a connection between meteors and comets. Two Italian astronomers took up this connection.

Giovanni Schiaparelli (1835–1910) is best known as a planetary observer, famously mistranslated as having seen "canals" on Mars, sparking hundreds of science fiction tales and a fair amount of observational wishful thinking. Among his major contributions to astronomy were detailed observations of the planets, several comets (including P/Swift–Tuttle in 1862), and meteors: Schiaparelli observed the Perseids in 1866, and the Bielid (Andromedid) storm of 1872. In his theoretical studies of meteors, Schiaparelli arrived at the view that there exist in the Solar System "currents" of meteoric debris, through which the Earth periodically runs – equivalent to what we now know as meteor streams. Schiaparelli corresponded on these ideas with Angelo Secchi (1818–78), who first published them on his behalf in 1866.

Secchi, too, was a noted observer. He made detailed observations of the Sun, several comets – notably P/Biela, with its split nucleus, in 1846 – and the planets. He was also interested in meteors, and arranged for simultaneous observations, similar to those by Brandes and Benzenberg, to be made from Rome and from Civitavécchia, 65 km (40 miles) away. On the basis of the results, Secchi calculated that meteors occur at altitudes of 75–250 km (45–150 miles). From detailed observations of the Perseid meteors, Schiaparelli concluded that they were produced by particles occupying similar orbits to the Great Comet of 1862, Comet P/Swift–Tuttle.

Schiaparelli gave a full account of these theories in his 1871 work *Entwurf einer astronomischen Theorie der Sternschnuppen* ("Outline of an Astronomical Theory of Shooting Stars"):

> The meteor showers are the product of the dissolution of the comets and consist of very minute particles that they … have abandoned along their orbit because of the disintegrating force that the Sun and the planets exert on the very fine matter of which they are composed.

The comet–meteor connection was cemented when other astronomers, including Kirkwood, identified further orbital similarities between shower-producing particles and particular comets. The Orionids and Eta Aquarids were found to be associated with Comet P/Halley, while the Leonid parent was identified as the short-period Comet P/Tempel–Tuttle.

The Leonids of 1866

Having been successfully predicted to return in 1833, the Leonids attracted still greater scientific interest at their next predicted appearance, in 1866. Hubert Anson Newton (1830–96) of Yale University had confidently forecast another storm. A study was also carried out by the mathematical astronomer John Couch Adams (1819–92), whose painstaking analysis of gravitational perturbations of the Leonid stream greatly improved the orbital elements. Once again the prediction was borne out, and a spectacular display was enjoyed by European observers. A detailed observing plan had been prepared in order to obtain as much information as possible about the Leonid meteors and their nature. Observations were made from Greenwich and elsewhere.

Among the observers was Alexander Stewart Herschel (1836–1907), perhaps the least known of the family of great astronomers. He made detailed measurements of the radiant position while observing the storm from Glasgow along with Professor Grant and his assistants from Anderson's College (now Strathclyde University). Herschel designed a hand-held binocular spectroscope for observing meteors. Such instruments were used, under the auspices of the British Association, by John Browning at Upper Holloway in London to observe Leonid spectra during the 1866 storm. Browning observed green lines in the spectra, interpreted at the time as being due to thallium (but more likely to have been due to atmospheric oxygen). Estimates of the peak rates produced during the 1866 Leonid storm over western Europe ranged from 2000 to 5000 meteors per hour.

Biela – the comet that vanished

If, by the late nineteenth century, proof were still needed of the relationship between comets and meteor showers, surely the most conclusive evidence was provided by the short history of Comet P/Biela. Like Comet P/Halley, this was an object named, not for its initial discoverer, but for the astronomer who first described its orbit. The comet was first found in 1772 by the French observer Jacques Leibax Montaigne. His countryman Jean Louis Pons (1761–1831), who still holds the record for the most cometary discoveries (37, though not all of these bear his name), found it again in 1805. Following the third discovery, by Wilhelm von Biela (1782–1856) in 1826, the identity of the comet with those of 1772 and 1805 was established.

Comet P/Biela had a short-period orbit of about six years, typical for a member of the Jupiter's family. The relatively short interval between successive perihelia took its toll on the comet, and at its 1845–6 return it was observed to have split into two. Gary W. Kronk suggests that the actual disruption may have occurred some years previously. The double comet returned in 1852, when it was observed by Secchi. Comet P/Biela was badly placed for observation in 1859, but was expected to be particularly favourable in 1865–6. Careful searches revealed nothing of the comet, which was presumed to have disintegrated completely.

Another return, bringing the comet close to Earth, would have been in 1872, but again it failed to show up. However, observers were surprised by a very fine meteor storm with its radiant near the star Gamma Andromedae on the night of 1872 November 27. Rates of the order of 6000 meteors per hour – fully comparable with good Leonid returns – were recorded. The display was observed by Schiaparelli and also by his colleague Secchi who, describing the event, wrote that "the layers of distant light resemble the surge of snowflakes." Secchi obtained spectra of the meteors which showed the presence of iron, magnesium, and sodium emission lines.

At what would have been a further return, on 1885 November 27, British observers reported peak rates of around 75,000 Andromedid meteors per hour. The storm was in progress at nightfall, as noted by the Scottish observer Smieton from Broughty Ferry near Dundee. Weaker displays were seen in 1899 and 1904, but no further really significant returns of the Andromedids have been seen since then. Perturbations by Jupiter have, for the time being, carried the stream's orbit away from its intersection with that of the Earth. As we saw in Chapter 1, computer simulations suggest that the Andromedids will return to the Earth's skies in the twenty-second century.

The late nineteenth century

By the late nineteenth century, many observers were taking up the challenge of systematic work, making measurements of meteor positions against the sky background and beginning to unravel the annual pattern of activity. The regular, dependable showers such as the Perseids came in for more attention. Alexander Herschel made observations of Perseid spectra, using a hand-held spectroscope; fast reactions were needed for this work, but he succeeded in sketching the principal meteoric emission lines.

Herschel was reputedly very accurate in plotting meteor paths, and did some work on computing meteor heights. With Robert Philips Greg (1826–1906), he prepared catalogues of meteor shower radiants. He was also involved in preparing reports on "luminous meteors" for various committees of the British Association from the 1860s to the 1880s.

The late nineteenth century has been seen as something of a golden age for amateur astronomy. Leading contributions to the science were made by "gentleman astronomers" – wealthy, leisured amateurs with the financial resources to establish well-equipped observatories and the time to use them. In England in particular, there were several noted planetary observers who fell into this category. Many of these private observatories were better equipped than the university establishments of the day. Several of the late-nineteenth-century amateurs took an interest in meteor observing, laying the foundations for their generally less leisured and substantially less wealthy modern successors.

Notable among the British observers of this time was William Frederick Denning (1848–1931) of Bristol, a remarkable character. Denning gave up a lucrative career as an accountant to devote his life to astronomical observing, living out his later years in abject poverty. He was certainly a prolific observer, discovering five comets and a nova, in addition to his meteor work. He also found time to write a popular book, *Telescopic Work for Starlight Evenings*.

A standard technique among meteor observers at this time was to plot observed meteor paths on gnomonic charts (see Chapter 4), specially prepared by the British Association. Denning used a slightly different approach, plotting his observations on celestial globes, from which the positions could be read later. During his long observing career he made thousands of such plots, which he used in preparing a detailed catalogue listing 278 radiants. This *General Catalogue of the Radiant Points of Meteoric Showers and of Fireballs and Shooting Stars Observed at more than one Station* was published as a Memoir of the Royal Astronomical Society in 1899. While many of Denning's catalogued radiants turned out to be spurious, he succeeded in identifying several showers, including the Taurids.

Among Denning's noted contemporaries was Thomas W. Backhouse (1842–1920) of Sunderland, who used similar observational techniques. He was responsible for preparing a set of 14 gnomonic charts for use in meteor plotting, based on a catalogue of almost 10,000 naked-eye stars. The charts remain in use to the present day.

The establishment in 1890 of the British Astronomical Association (BAA) provided an important focus for amateur astronomers. The BAA's various observational sections, each devoted to specific areas such as solar, lunar, planetary, variable-star and, of course, meteor observing, continue to the present day as internationally respected clearing-houses for amateur work. The coordinated efforts of the BAA Meteor Section over the years have led to the collection of a vast archive of data from observers around the world. The continuity of these results is of great value in long-term studies of how meteor streams behave – only by examining directly comparable data sets from many years, or even decades, can subtle changes in activity pattern be detected.

The twentieth century

In parallel with developments in Great Britain, the American Meteor Society was established in 1911. It did much valuable work, coordinating meteor observations in the United States and Canada under the guidance of the professional meteor astronomer Charles P. Olivier (1884–1975). Substantial sections of Olivier's classic text, *Meteors*, are based on results obtained by amateur observers.

The value of meteor observations by amateur astronomers has always been recognized by the professionals. Since the beginning of the twentieth century, most professional astronomical research seems to have been directed towards cosmological, astrophysical, or planetary studies, areas which require the use of large telescopes. Meteor studies have little need of such equipment: the naked eye and the wide-field camera remain two of the best instruments available. Few professional astronomers, however, have the time to devote to meteor observing, which (as I shall outline in subsequent chapters) demands a certain degree of patience, and, at times, a long wait before sufficient data can be gathered. The interested amateur is generally more prepared to put in this sort of time than the professional theoretician.

Some professional astronomers working on meteoric problems have put in their fair share of observing time. During the 1930s, the German astronomer Cuno Hoffmeister devoted considerable effort to visual meteor plotting and the determination of radiants. He mounted observing expeditions to cover radiants in the southern sky, and compiled a radiant catalogue (*Meteorströme*) based on over 5000 German observations.

Hoffmeister believed that a substantial proportion of meteors were produced by particles in hyperbolic paths, which would mean that they could originate from beyond the Solar System. Another proponent of this view was Ernst Julius Öpik (1893–1985). As a research associate at Harvard University, Öpik undertook an observing expedition to Arizona in 1931–3 in order to obtain meteor photographs for measurement. His cameras were equipped with a rocking-mirror system which produced curved meteor images on the photographic plate. Meteor velocities were determined by measuring the curved trails, and the results yielded values apparently consistent with hyperbolic meteoroid trajectories. The possibility of an extra-solar origin for meteoroids was not finally ruled out until the advent of more sophisticated photographic and radio meteor observation techniques in the 1950s, and was to remain a central area of contention among professional meteor astronomers in the years between the wars.

Also at Harvard at this time was Fred Whipple, who later became noted for his studies of comets. Whipple spent many years coordinating photographic meteor patrol programmes, with the aim of recording trails for accurate positional and other work. The Harvard Meteor Project, conducted from two observing sites in Massachussets from 1936 to 1942, and later from New Mexico, yielded a great deal of useful data on the movement and activity of shower radiants, and rates for brighter meteors. Important among these studies were observations of the Geminids and Taurids, made using cameras driven to follow the stars, and equipped with rotating shutters ("choppers") to aid meteor identification.

Another prolific professional student of meteors at this time was the Canadian Peter Millman, who carried out pioneering work on photographic meteor spectroscopy from Harvard in the early 1930s. Detailed examination of such spectra allowed something of the chemical nature of meteoroids to be determined. After the Second World War, Millman conducted photographic observations from the Dominion Observatory at Ottawa. Among his noted successes was photography of the Giacobinid meteor storm on 1946 October 9–10. Later, during the highly successful International Geophysical Year, organized during 1957–8, Millman was responsible for coordinating a worldwide programme of visual meteor observations.

Radar and radio observations

R adio astronomy was still in its infancy when its applicability to meteor studies was revealed by Bernard Lovell at Jodrell Bank in the 1940s. During the Second World War, radar operators detected the occasional spurious echo which could not be accounted for by aircraft. Some researchers thought that these might mark the arrival of energetic atomic particles from space – cosmic rays. In 1946, equipped with war-surplus radar equipment, Lovell and his colleagues at the University of Manchester set out to address this question. He had a suspicion that meteors, rather than cosmic rays, might be the cause of these echoes.

Lovell was certainly quick to recognize the value of amateur observers in assisting with his work. Later, describing the efforts of the BAA Meteor Section in his 1968 book, *The Story of Jodrell Bank*, he acknowledged that "The published works of these people and their predecessors were priceless contributions to astronomy."

In his pioneering radar observations of meteors, Lovell hoped to detect reflections from the brief trails of ionization produced by meteors in the high atmosphere. To do this effectively, however, he had to call on amateur observers to provide visual confirmation of events. The then Director of the BAA Meteor Section, J.P.M. Prentice, a very distinguished amateur meteor observer and a solicitor by profession, lived in Stowmarket, Suffolk, but willingly travelled to Cheshire in August 1946 to visually augment the Jodrell Bank attempts to observe that year's Perseids by radar.

The results, unfortunately, were inconclusive, but Prentice did alert Lovell's team to the possibility of strong meteor activity that autumn from the Giacobinids – a swarm of debris close to the nucleus of Comet P/Giacobini–Zinner which, like the Leonids, can sometimes produce storm levels of activity. On the night of 1946 November 9–10, the radar equipment was again switched on. Few echoes were obtained above the normal background rate until about midnight when, suddenly, the echo rate rose dramatically. A quick visual check revealed that a Giacobinid storm was in progress, with hundreds of meteors visible every minute. The peak came in the early morning hours. These observations finally confirmed the usefulness of radio and radar as means of detecting meteor activity. Across the Atlantic, the same event was photographed from Canada by Millman.

During the Giacobinid storm, the Jodrell Bank observers found that strong radar echoes were picked up only when the antenna was aimed in a direction 90° in azimuth away from the shower radiant in Draco. At last, an explanation was found for the lack of correlations between visual sightings and radar echoes during the previous summer's Perseid observing campaign.

Further fruitful radar and radio observations were carried out at Jodrell Bank in the late 1940s and 1950s. Notable among this work was the discovery of daytime meteor showers, which – for obvious reasons – could not previously be observed. During June and July, strong radio meteor activity is produced by a number of streams, including a Beta Taurid shower, which is the result of an encounter between the Earth and the same stream that produces the nighttime visual Taurids of November.

Prentice remained a regular collaborator with the Jodrell Bank radio meteor workers, and was later awarded an honorary masters degree by the University of Manchester in recognition of his valuable contributions to the project.

Professional meteor studies by radio and radar are now carried out on a regular basis mainly from Japan and the countries of the former Soviet Union. Occasional special projects have been mounted elsewhere. A notable example was the radio meteor project conducted by Zdeněk Šekanina during the 1960s, which aimed to assess the day-to-day meteoroid flux and detect new daylight streams. Other occasional projects have included work coordinated from Sheffield University in which radar systems were operated from the Outer Hebrides off the coast of Scotland in the 1970s to obtain meteor reflections as a means of investigating high-atmosphere winds.

Meteor photography programmes: the super-Schmidts

The flux of meteoroidal material arriving in near-Earth space became of increasing interest as the space age dawned in the late 1950s. It became important for those planning space missions to know what the chances were of spacecraft – particularly manned spacecraft – being damaged by meteoroid impacts. To this end, several important lines of investigation were set up.

The value of photography for recording meteors had been appreciated during the Harvard Projects led by Whipple in the 1930s and 1940s. Improvements in photographic equipment and films made it possible to record fainter meteors, and to assess the small particle flux more accurately. In particular, the super-Schmidt cameras designed by James G. Baker in the 1950s made photography down to the fainter end of the magnitude range a matter of routine.

The Schmidt camera has been invaluable in astrophotography. Its particular advantages are "fast" optics – allowing rapid recording of faint objects – and a relatively wide field of view. The system, as adapted for meteor photography, provided a 55° field of view, and used an aperture of 310 mm (12¼ inches). Meteors down to about fourth magnitude could be recorded, with an average capture rate of 2 meteors per hour. A rotating-shutter mechanism, fed through a central hole in the mirror, allowed meteor trails to be easily recognized. The super-Schmidts used for meteor work were equatorially driven to follow the stars during exposure, providing point images for positional measurement.

Harvard College Observatory operated four super-Schmidt cameras from sites in New Mexico, and two more were run at sites in Alberta, Canada, by Millman as part of a project for the Dominion Observatory. The outcome of this work was a detailed catalogue from Harvard of meteor orbits and velocities. From this work, and the radio meteor observations at Jodrell Bank, it was conclusively shown that meteoroids originate exclusively within the Solar System, finally laying to rest the hypothesis that some meteors could result from the arrival of particles on hyperbolic trajectories from interstellar space. It was also possible to deduce from this work, and from studies using meteoroid detectors on board artificial satellites, that the numbers of large meteoroids traversing near-Earth space were sufficiently small to allow manned space missions both in Earth orbit and, more significantly, to the Moon, to proceed in relative safety.

The 1966 Leonids

As we have already seen, observations (both chance and planned) of the periodic storms which can mark the perihelion passage of Comet P/Tempel–Tuttle have done much to improve our understanding of meteors. The 1866 return was intensely observed. The expected displays in 1899 and 1933 were, in the end, disappointing, and few astronomers had high hopes for the 1966 return.

Amateur observations in the years leading up to 1966 showed some increase in activity. The then Director of the BAA Meteor Section, Harold Ridley, conducted a study of past observations. Unlike many others, Ridley was optimistic about the chances of good activity, as published in a *Sky & Telescope* article. His forecasts bore spectacular fruit on the night of 1966 November 16–17.

Observers in Britain conducted night-long vigils in the hope of seeing something, but found only moderate rates – though in terms of the Leonids in a good year, even "moderate" rates correspond to a respectable Perseid return. The great storms of the Leonids are confined to a fairly short time interval, and are therefore observed only from limited longitudes on the Earth. European observers in 1966 missed out by just a few hours. Their colleagues in America, on the other hand, were treated, for a period of about 40 minutes around 1155 UT (see p. 68), to peak rates of 60,000 meteors per hour – better even than the 1799, 1833, and 1866 displays.

In America, observations were made from Kitt Peak, where Dennis Milon took a number of photographs of the display. On exposures lasting only a couple of minutes, Milon recorded dozens of meteors – these only the brightest of the crop. The radiant effect is clearly shown, and on one frame, two meteors coming head-on towards the observer out of the radiant itself, in the Sickle of Leo, were captured.

Amateur observations

The valuable assistance given by J.P.M. Prentice in the professional radio work conducted at Jodrell Bank has already been mentioned. At this time, Prentice was in charge of a thriving BAA Meteor Section whose active observers obtained several other important results. A mainstay of the BAA observing programme at this time remained the plotting of meteor paths for positional analysis.

Prentice's successor, Harold Ridley, distinguished himself in the field of meteor photography and spectroscopy during the 1950s, at a time when such work was still largely beyond the scope of all but the most ingenious amateurs. Ridley's successes included a superb, detailed Taurid spectrum photographed in 1954, and high-quality direct photographs which proved valuable in measuring the precise positions of several meteor shower radiants. With his encouragement, the Lloyd-Evans brothers (Tom and Robert) went on to emulate Ridley's success from a dark site near Forfar in Scotland during the early 1960s. Among their notable results was the spectrum of a Lyrid meteor, recorded in 1960, showing 70 lines. At one time, the BAA observers' contributions constituted a significant fraction of all meteor spectra recorded worldwide.

Visual and photographic observations both continue to play a large part in the activities of groups such as the BAA Meteor Section. Visual observing techniques (described in Chapter 4) concentrated largely on meteor rates and magnitudes from the 1960s onwards, leaving positional work to the photographic observers.

The return of Comet P/Halley in 1985–6 was the cue for concerted professional and amateur observation programmes. Again, meteor observers had their part to play in what became known as the International Halley Watch. Observations were collected from around the world of the Eta Aquarid and Orionid showers associated with the comet.

Fireball patrols

Several joint amateur–professional programmes for fireball patrol photography using wide-field lenses were also instigated during the 1960s in the hope of recording the bright events associated with meteorite falls. Among the successful patrols was the Prairie Network, which operated from 16 stations in the American Midwest between 1964 and 1975. The cameras at these stations were run by farmers on a voluntary basis, and were simple to operate. Many valuable results on the flux of fireball-producing objects were obtained. The collection of the Lost City Meteorite, following analysis of patrol photographs of a very bright fireball over Oklahoma in 1970, was a major success for the network.

Other, similar networks operated elsewhere. In 1977 patrol photography in Canada resulted in the collection of the Innisfree Meteorite. A Czechoslovakian network consisting of 20 stations, established in 1964, also enjoyed some success following the boost to meteor photography given by the recovery of the Příbram Meteorite in 1959. Rather less success, unfortunately, was enjoyed by the amateur network operated in Britain by the BAA in the early 1970s. Sadly, although there is still considerable value in such work, none of these networks survived far into the 1980s.

In the 1980s, amateur astronomers around the world continued to be the main gatherers of observational meteor data. Successful groups are currently run in several countries including Japan, the former Soviet Union, the Netherlands, the UK and Ireland, and the United States. National organizations provide the main coordinating base for this work. The BAA Meteor Section continues to flourish, and maintains excellent contacts with professional meteor workers, as do its Japanese, Russian, and European counterparts. In addition to making their results available for professional use, these organizations keep in regular, informal contact with each other. In the late 1980s, an International Meteor Organization was formed by Belgian amateurs, with the aim of pooling observational work from amateurs across the whole globe in order to obtain uninterrupted coverage of the major showers.

Professional astronomers nowadays have less and less time for the nuts and bolts of observing, and the amateur, with his or her time and enthusiasm, has an increasing role to play. It is interesting to note, as a parallel to professional interest in amateurs' meteor work, the enthusiasm with which many astrophysicists welcome amateur observations of variable stars. Similarly, amateur observers have frequently been responsible for making – and rapidly reporting – comet, nova, and other discoveries which have resulted in important professional observations being made.

As early as the late 1980s, professional workers were beginning to establish contacts with a view to collecting amateur observations of the expected strong display – maybe even a storm – of Leonids at the end of the twentieth century. Sustained observations, made in a consistent manner, of the kind which amateurs have been collecting for years, are proving of great interest to professional students of this shower's behaviour. A number of professional-led amateur observing projects will probably be set up to cover this exciting event, as happened with Comet P/Halley during its 1985–6 return.

The Leonids will doubtless receive (literally!) the lion's share of media and professional attention in the coming years, but there are other areas where professionals and amateurs with a common interest in meteors will surely interact. Further six-yearly Giacobinid storms are always possible. There is renewed interest in the Perseids now that their parent comet (P/Swift–Tuttle) has finally returned from its wanderings. The Geminids continue to excite much theoretical interest as an apparently rapidly evolving stream: amateur observations will certainly be sought in attempts to prove, or disprove, existing models.

Many amateur astronomers are now using photographic, CCD (charge-coupled device), and even radio equipment which represented the professional state of the art only a few decades or even a few years ago. The application of such technical advances means that amateurs will continue to add to our understanding of meteors and related phenomena well into the future.

The future

CHAPTER 4

Naked-eye visual observation of meteors

Meteor plotting

Meteor observation remains one of the most accessible areas for the amateur astronomer, and one in which useful contributions may still be made using a bare minimum of equipment. However, the observational methods used to obtain scientifically worthwhile meteor results have changed somewhat over the years.

For a long time until the late 1950s, a mainstay of amateur observation was the plotting of meteor paths on specially prepared *gnomonic charts*. Meteors appear to travel along short arcs of great circles on the celestial sphere. More conventional star charts, such as those to be found in the ubiquitous *Norton's* star atlas, use projections unsuitable for accurate meteor plotting – on such charts meteors would have to be plotted as curves. Gnomonic charts get round this drawback by using map projections centred on several different parts of the sky. By selecting the appropriate chart, most meteors can be plotted as straight lines with reasonable accuracy. At larger distances from a chart's centre, however, distortion of the plotted star groupings can be a problem, and it is sometimes necessary to use a number of overlapping charts.

Plotting was used for a long time to determine accurate positions of meteor shower radiants, while combined plots of the same meteor as seen by two or more widely separated observers could also be used to triangulate meteor heights and positions in the atmosphere, as was attempted as early as 1799 by the German students Brandes and Benzenberg. Particularly prominent for their efforts in this area were the English observers J.P.M. Prentice and George Alcock, who became well known for his many discoveries of comets and novae.

Alcock has related how his enthusiasm for meteor plotting was dampened by an unfortunate instance of a bright meteor which should have been very well plotted, but was not. Both he and Prentice had been observing one November night from their respective locations at Peterborough and Stowmarket, but Prentice stopped a few minutes earlier than Alcock, and so missed the fireball.

This frustrating episode highlights one of the problems of visual meteor plotting: if the meteor is plotted from only one station, triangulation is impossible. Variations in observers' perceptions may mean that two observers cooperating to attempt triangulation by looking towards the same part of the atmosphere seldom see the same event. Observers also vary in their ability to make accurate plots of meteors, sometimes introducing considerable errors.

Few amateurs now plot meteor trails, as trajectories may be much more accurately recorded photographically, or by professional obser-vatories employing radar techniques. Instead, emphasis is placed on accurately determining rates and magnitudes, which can reveal much about the structure of meteor streams and about the particles within them.

Nothing could be simpler than naked-eye meteor observing: all the observer needs to do is watch an area of sky for a reasonable period of time, and record the details of any meteors seen. Naked-eye meteor observation exploits the human eye's natural wide field of view. While very easy to carry out, meteor watches, like any other form of observational activity, do require some degree of advance planning.

Planning a meteor watch

When to observe

There is little point in attempting long watches on nights when the Moon is bright and filling the sky with its glare, since all but the brightest meteors will be lost against the sky background. Haze has a similarly adverse effect on observed meteor rates.

It must also be said that novice meteor observers will find little to interest them initially on those nights when major shower activity is absent. The low-activity nights of early spring are not the best time to start out in meteor observing, as disillusionment will rapidly set in. Low rates should not, however, deter experienced observers, who need no reminder that their results from such nights are every bit as valuable as those from the times of year when the pickings are richer.

Most meteor observers have their introduction to the field during one of the more active showers of the year. The Perseids in early to mid-August are a great favourite, since they produce consistently high rates and many bright meteors, and occur at a time of year when weather conditions in North America and Europe are often very pleasant for observing. Good rates are also produced by the Geminids, though observing conditions in northwest Europe tend to be poor during the mid-December maximum, and the low temperatures on nights which are clear at this time of year are not very welcoming for the newcomer.

Meteor watches may be carried out by solitary observers, or by groups of observers. There are many who enjoy the tranquillity of observing alone into the small, quiet hours of the night, but company can lift morale during intervals when rates are low, and stimulate greater effort on occasion: observers taking part in group watches tend to give up less easily when, for example, patches of cloud appear.

Where to observe

Choice of observing location is, of course, important. In common with most astronomical activities, meteor observing requires dark unobstructed skies, and it is desirable to find a site well away from artifical lights. If a dark garden can be used, there will be no problem, but observers based in built-up areas may have to be prepared to travel to escape the all-pervasive light pollution. Again, there are benefits in this respect in observing as part of a group within a local astronomical society.

The chances are that within a local society at least a couple of members will have access to dark observing sites, and will be prepared to share them with others. Alternatively, a group of like-minded observers in a society may be prepared to share the costs and provide the transport for occasional trips to dark locations. Some local astronomical societies have their own observatories at dark sites, which have the added advantage of acting as a base for the watch and a shelter should the weather change for the worse.

Obviously, one should avoid trespassing on, or causing damage to, property and farmland in the search for a dark observing site. Permission should always be sought before setting up to observe on an apparently unwanted piece of land. If you are going to a remote location to observe, it is advisable to take a travel first-aid kit, and to let someone know of your expected movements, especially if you are going alone.

What you will need

Naked-eye meteor observation requires little more equipment than the eye itself! There are a few other essentials, though, which meteor observing shares with most other areas of astronomy.

Seating

The meteor observer will ideally spend much time looking up into the night sky, in directions about 50° above the horizon. Standing around doing this for any length of time is an excellent way of developing neck strain, and should therefore be avoided. Comfort is essential to the observer's concentration, and an observer with a cricked neck is likely to miss a fair proportion of meteors. The ideal solution is to recline in a deckchair or on a sun-lounger or an inflatable airbed.

Clothing

Warmth is another essential aspect of observer comfort. A cold observer is an unreliable observer. Even on summer nights, the damp can lead to cold feet long before the dawn, so wrap up well before starting to observe. It is worth remembering is that it is easier to stay warm than to warm up after becoming cold. If necessary, overdress before starting to observe: putting on an extra pullover will do little to warm you up once you have been out in the cold for a couple of hours. Much body heat is lost through the feet so wear an extra pair of socks. Keep your head covered, too. A rug can be used to cover your legs. A warm parka is a good investment for winter meteor watches.

Some observers solve the problem of staying warm by climbing inside a sleeping-bag. If the observer has to attend to cameras or other equipment in addition to carrying out the visual watch, however, it can become a nuisance to have to climb in and out of a sleeping-bag repeatedly during the session.

The importance of staying warm during the watch, especially on those crisp, transparent winter nights which provide the best skies, cannot be overemphasized. On a number of occasions I have acquired a mantle of frost from my condensed breath over my outer clothing during December and January meteor watches. Fortunately, adequate layers of warm clothing beneath the outer layer prevented me from becoming too cold to enjoy the Geminid or Quadrantid activity on those nights.

Astronomical essentials

There are a few items of simple equipment which are essential for meteor observation. Amateur astronomers who carry out observations of any form will already have, or certainly will require, these.

Timepiece Since an important part of meteor observing is the accurate recording of times, you should ensure that you have a good, reliable watch, adjusted to the correct time. Broadcast or telephone time signals can be used to ensure that the watch is accurately set. Cheap digital watches are readily available, and allow the time to be rapidly read. Avoid using watches with illuminated displays, however, since these will destroy the dark adaption (see below) of the eye which is essential to meteor observing.

Clipboard and paper Use a clipboard to hold the sheets on which your results are to be recorded. Make sure that it holds the paper firmly – there is nothing more disruptive to a meteor watch than having to chase the previous half-hour's results across a dark muddy field because they have been blown away by the wind! Keep a plentiful supply of paper to hand. Many observers write their results in a rough shorthand during the watch, for later transcription onto proper report sheets. There is considerable merit to this system, as it helps to reduce "dead time" – time spent looking down, rather than at the sky.

Red torch/flashlight On leaving a brightly lit room to go out observing, you will find that it takes some time for the fainter stars to become apparent to your vision. This change to night vision, called *dark adaption*, is a physiological process which takes about 20 minutes to become complete. Without it, only the brightest meteors can be seen. Obviously, the observer does not want to have to repeat the process every time a meteor is seen by dazzling himself or herself while writing down the details by the light of an ordinary torch/flashlight. One with a dim red light is used instead, as this has little effect on dark adaption. Rear bicycle lamps may be used, though these can still be too bright. My own solution, which has worked well over the years, has been to use an ordinary torch/flashlight with the glass covered by overlapping strips of red tape.

Spare batteries Spode's law, which states that if anything can go wrong, it will, and is well known to amateur astronomers, implies that your torch/flashlight batteries will run out on Perseid maximum just as things are getting interesting, so that you will be unable to make useful observations or even keep a sensible record. Carry spares!

Pens and pencils Spode also attacks these simple pieces of equipment – pens gum up in the cold and damp, while pencil points frequently break. Be prepared.

Star atlas It is always useful to have a star atlas handy as an aid to identifying shower radiant positions. *Norton's*, a long-established favourite of meteor observers, shows stars to just below the naked-eye limit.

Refreshments During long watches, the observer may become hungry and thirsty. While alcohol should, for obvious reasons, be avoided, a hot drink and a light snack may be very welcome during long observing stints.

Tape recorder – or not? Some authorities suggest recording observations for later transcription. I tend to err on the side of caution, and advise against the use of portable cassette-recorders, especially in cold or damp conditions. Several observers (some of them very experienced) have lost valuable results to battery or mechanical failure.

Patience The most important component in the meteor observer's toolkit is the mental attribute of patience. On occasion, long spells will pass without any meteors being seen, particularly on those nights when there is little shower activity. The observer must maintain concentration, however, or meteors will be missed, further lengthening the gap between sightings.

Carrying out the watch

Where to look

Before starting to observe on any night, it is important to know which showers are expected to be active. This can be checked by consulting lists such as Table 5.1 (pp. 86–7), or those published annually in the handbooks of the BAA and RASC, or in *Norton's 2000.0* star atlas. These lists will give the expected position for each radiant on its night of maximum activity. If observing on nights away from the maximum, remember to allow for the radiant's drift across the sky, resulting from the changing aspect of the Earth with respect to the meteor stream as it continues around its orbit. As an observational convenience the radiant is taken to lie at the centre of an 8° diameter circle, from which shower members might be expected to emanate (this does not necessarily reflect any real property of the stream). It is useful to plot the radiant position on a suitable star map well before starting the watch, and taking a little time at the beginning of the watch to become familiar with its position.

In carrying out the watch, it is of course important to look in the right direction. During showers such as the Perseids, many observers make the mistake of looking straight at the radiant, but most meteors from a given shower are actually seen some distance away, for a number of reasons. As amply demonstrated during the great Leonid storms, meteors close to a radiant appear short – sometimes even as point sources – and these the observer is likely to miss. The human eye is well suited to detecting rapid movement, however, and the longer streaks produced by meteors at a reasonable distance from the radiant are actually easier to detect. Consequently, most meteors are seen about 40° from the radiant.

The elevation towards which the observer looks is also important. Remember that meteors appear in a volume of the Earth's atmosphere above the observer, so, in theory, the maximum number might be expected if the observer's field of view could take in the greatest possible volume of atmosphere, just above the horizon. Overhead, the field of

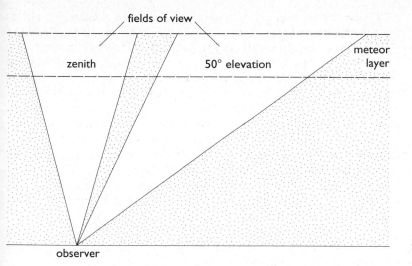

Figure 4.1 *An observer will, potentially, see many more meteors by looking at 50° above the horizon, than by looking at the zenith: at 50°, a greater "wedge" of atmosphere is covered by the field of view.*

view encompasses only a small volume of atmosphere, and consequently a much smaller number of potential meteor sightings. In practice, a compromise is best. Looking through a greater volume of atmosphere also means looking through more intervening atmospheric grime, so many meteors are lost in the background. At 50° elevation, however, the observer should be looking through clearer air and a reasonably large volume of atmosphere (Figure 4.1), in which many more meteors will be seen than at the zenith.

It is therefore recommended that observations are centred on an area of sky 50° above the horizon and 40° to one side of an active radiant. On nights when sporadic activity is the sole meteoric attraction, the observer should look to the east or northeast.

Duration

Meteor results are usually analysed to provide hourly rates, and meteor watches should therefore ideally be at least an hour long. It should be possible on a good night to obtain several hours' data, but it must be stressed that the observer's attention tends to drift after a while, and it is more sensible to do stints of, say, 2 hours on, 15 minutes off, alternating for as long as you wish to continue. Breaks can be used to take light refreshments and help to maintain alertness.

Meteor observing is not a competitive sport, and there is no point whatsoever in aiming for record-length watches – in practice, the theoretically possible 14-hour winter marathons are of little value after the first few hours, as observer fatigue will reduce the validity of the later hours' results. During the pleasant, warmer summer nights, watches of 4 or 5 hours' unbroken duration are just about feasible, and make full use of the short period of darkness.

What to record: general details

The organizations which collect visual meteor reports usually request a few items of background information which help, as we shall see later, in the subsequent analysis of the observations. Space for recording these is provided on the standard report forms on which the meteors seen during a night's observing will eventually be logged.

Date

Obviously, it is important to note the date and time of your observations. To avoid ambiguities which may arise in watches spanning midnight, meteor observers always use a double date in the form of, for example, 1990 September 25–26, meaning the evening of September 25th to the morning of September 26th. By convention, the night of, for example, July 31st to August 1st, 1992 is written as 1992 July 31–32.

Watch times

The start and end times of the watch must be recorded if rates during the watch period are to be determined. As for all astronomical timings, Universal Time (UT) should be used in meteor reports; 9.00 pm, for example, is written as 2100 UT. Universal Time is equivalent to Greenwich Mean Time, so observers in the UK should remember to subtract an hour from their timings when British Summer Time is in force. Observers in North America will be several hours behind UT, depending on their precise longitude, and should correct their timings accordingly. Table 4.1 gives the necessary corrections which should be applied to arrive at UT in the various US time zones.

Sky conditions

The sky conditions during a meteor watch will influence observed rates in a number of ways. Haze or moonlight, as mentioned earlier, will drown out the fainter meteors, and when the watch results are eventually analysed it is important to be able to allow for these. An index of sky transparency and darkness is given by the *limiting magnitude* (LM). This is basically the brightness of the *faintest* star visible to the observer. If faint stars are invisible, then faint meteors certainly will not be seen either. In practice, the faintest meteors seen during a watch are generally a magnitude or so brighter than the faintest stars.

Limiting magnitude may be estimated in a couple of ways, each of which has its adherents among meteor observers. The *north polar sequence* (Figure 4.2), a series of stars of accurately known magnitude scattered around Polaris, has been used by northern hemisphere observers for many years. Comparison stars in convenient variable-star fields (but *not* the variable stars themselves) may also be used. An alternative method is

---

Table 4.1 *United States time zones.*

Eastern Standard Time (EST)	UT minus 5 hours
Central Standard Time (CST)	UT minus 6 hours
Mountain Standard Time (MST)	UT minus 7 hours
Pacific Standard Time (PST)	UT minus 8 hours

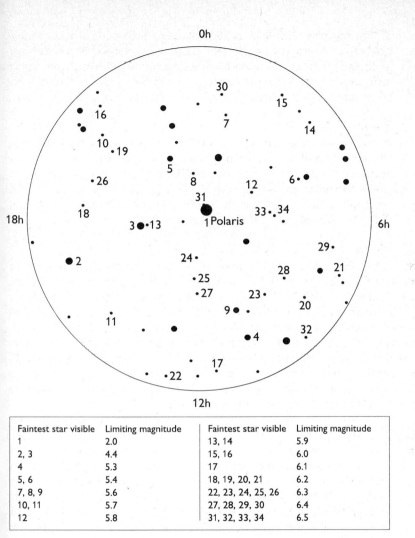

Faintest star visible	Limiting magnitude	Faintest star visible	Limiting magnitude
1	2.0	13, 14	5.9
2, 3	4.4	15, 16	6.0
4	5.3	17	6.1
5, 6	5.4	18, 19, 20, 21	6.2
7, 8, 9	5.6	22, 23, 24, 25, 26	6.3
10, 11	5.7	27, 28, 29, 30	6.4
12	5.8	31, 32, 33, 34	6.5

Figure 4.2 *The north polar sequence, selected stars within 10° of the north celestial pole used as a standard for estimating limiting magnitudes. It provides an extremely useful means of assessing sky transparency, and therefore of the likelihood of observing fainter meteors.*

to count the number of stars visible in small defined areas of sky bounded by bright stars. The method appears to have some limitations, however, when applied to areas of sky traversed by the Milky Way. Useful watches may be made when the sky limiting magnitude is better than 5.0.

Cloud also adversely affects meteor observations. Small amounts can be tolerated for limited periods during a watch, but there is little point in trying to spot meteors through the gaps in a sky 50 percent overcast for any length of time. Should cloud appear during a watch, estimates of its extent in tenths of the sky should be made at regular intervals and logged on the report sheet.

**What to record for
each meteor**

Once dark-adapted and settled down for the watch, the observer will hopefully begin to see meteors appearing occasionally in the field of view. For each meteor, there are a number of useful details which may be recorded.

Time of appearance

If rates are to be assessed for a given meteor shower, then a record of the time of appearance for each meteor is required. Usually it is adequate to record this time to the nearest minute of UT. Observers carrying out visual watches to complement photographic work may find it necessary to record times accurate to 6 seconds (0.1 minute) for the brighter objects.

Type

Given prior knowledge of activity on the night of observation, the observer may try to decide whether a particular meteor belongs to a shower known to be active. If a meteor's path can be extended backwards to

Figure 4.3 A known radiant position is used to determine whether meteors belong to a given shower. Plotted here as a circle is the Perseid radiant. The paths of meteors A and B can be traced back to within the radiant, so they can be assumed to be Perseids. Meteors C and D, however, have paths which cannot be extended back to the radiant, and are therefore sporadics.

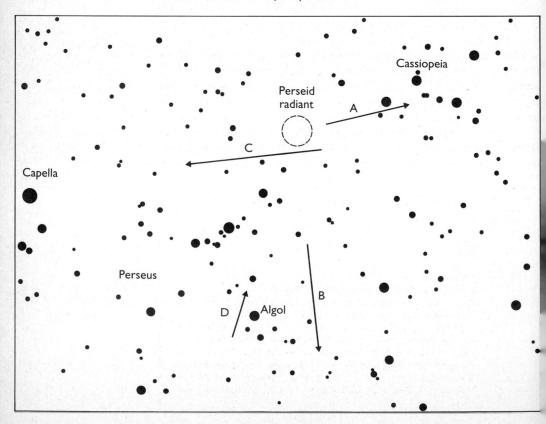

intersect an 8° diameter circle centred on the radiant of a shower active that night, then the meteor can be ascribed to that shower. Those meteors whose paths cannot be traced back to any known shower radiant are classed as sporadic, and belong to the random background population. For example, meteors A and B in Figure 4.3 are Perseids, while C and D are sporadics.

Magnitude

Many beginners find the estimation of meteor magnitudes a daunting prospect, and there are some who shy away from doing so altogether. But this is one of the most useful pieces of information which can be recorded. Accurate estimation of meteor brightness is one of several observational skills that will come with practice. If you make the effort during an active shower such as the Perseids, when there are plenty of chances to get the hang of the system, estimation of meteor magnitudes will soon become second nature.

As with stars, meteors are ranked in magnitude ranging from the faintest naked-eye range (about magnitude 5), through bright objects (magnitude 0), to very bright objects (fireballs, brighter than magnitude −5, and quite rare). Background stars and other objects provide suitable "comparisons" for the estimation of meteor magnitudes. For example, a meteor appearing as bright as the planet Jupiter is of magnitude −2, while one about the same brightness as the star Delta Ursae Majoris – the faint member of the Plough (Big Dipper) – is of magnitude 3. Table 4.2 provides a list of some suitable comparison objects.

A variation on this idea employed by some amateur groups, such as the UK's South Downs Astronomical Society, requires the observer simply to write down the identity of a star of similar magnitude to the meteor. This demands use of a star atlas, and, in the subsequent analysis, access to a star catalogue listing magnitudes.

In practice it is adequate to estimate a meteor's magnitude to the nearest whole magnitude. Observers of variable stars, whose targets are present in the field of view for rather longer than the typical

Table 4.2 *Some useful comparison objects for estimating meteor magnitudes.*

Magnitude	Object(s)
−4	Venus
−2	Jupiter
−1	Sirius
0	Capella, Arcturus, Vega, Rigel
1	Altair, Deneb, Aldebaran, Procyon, Pollux, Regulus
2	Polaris, Alpha and Beta Ursae Majoris (the Pointers), Alpha Cephei, Alpha Persei, Alpha Andromedae, Beta Aurigae, the Belt stars of Orion, Denebola, Gamma Draconis, Alpha Ophiuchi, Gamma Cygni, Epsilon Cygni
3	Albireo, Delta Ursae Majoris, Beta Arietis, Eta Boötis, Delta Herculis, Epsilon Boötis, Beta Canis Minoris, Gamma Aquilae, Delta Cygni
4	Eta Persei, Delta Aurigae, Nu Persei, Kappa Persei, Iota Herculis, Epsilon Aquilae, Beta Aquilae, Rho Boötis, Sigma Boötis, Rho Leonis, Gamma Trianguli
5	"Forepaws" of Ursa Major, Epsilon Lyrae. In practice, the faintest meteors visible to the naked eye

two-tenths of a second for the average meteor, claim accuracies of no better than 0.2 magnitude, and it would be slightly ludicrous for meteor observers to pretend that they could achieve similar accuracy.

Persistent trains

A number of meteors, particularly the brighter ones, may leave behind them a briefly luminous persistent train resulting from ionization of the upper atmosphere during the meteor's passage. The train appears as a brief streak of fading light, remaining for a short period after the meteor itself has vanished. Typically, such trains last for no more than a few seconds, but sometimes durations of over 30 seconds, and even up to several minutes, may be recorded.

Careful recording of trains, coupled with magnitude data, can reveal something about the nature of particles within a meteor stream. Perseids and Orionids are regarded as classic examples of meteors derived from cometary sources, and are noted for their high frequency of persistent trains. Geminid meteors, on the other hand, are believed to be produced by particles of asteroidal origin, and so to have greater structural integrity than cometary material. The relative dearth of Geminids leaving persistent trains appears to point to this different origin.

The velocity of the incoming meteoroid as it strikes the upper atmosphere is also important in determining whether the collision between the particle and the atmosphere releases sufficient energy to produce ionization. Showers producing slow meteors, such as the Taurids, provide relatively few trains, while Leonids, which enter the atmosphere at velocities close to the theoretical maximum, often produce trains.

The state of the atmosphere also plays a role. During Perseid watches I have often noted that meteors appearing in that part of the sky towards the twilight arch tend to leave slightly more persistent trains, and of longer duration, than those appearing in the dark half of the sky. This region of the meteor layer perhaps retains more of the daytime excitation produced by the action of sunlight, meaning that a greater proportion of atmospheric molecules are already "primed" at higher energies before being further excited by meteor-related ionization. Consequently, such atmospheric particles remain ionized for slightly longer, following the passage of a meteor, than those in the darker half of the sky, which have had more time to de-excite.

It is usual to record the duration (in seconds) of a persistent train. One method is for the observer to "count thousands" – each second being timed by mentally saying "one thousand" (one second), "two thousand" (two seconds), and so on. Sometimes trains will be of such extremely short duration (less than 0.5 second) that their lifetimes cannot be accurately estimated by the observer. Such trains are technically described as *wakes*.

The behaviour of long-duration trains may be interesting to watch, revealing as it does the action of high atmospheric winds. After a short time a train may become quite markedly distorted by winds in the meteor layer, which can have velocities of around 400 km/h (250 mile/h). Sketches of rare long-duration events can be of interest; binoculars are useful to reveal more detail.

Reliability

The accuracy with which the observer feels that the details have been recorded is given on the report forms used by some organizations on a three-point scale – A for very good, B for average, C for poor. Typically, the meteors for which the accuracy of recording will be best are those which appear directly in the observer's field of view and are well seen, while it is harder to record accurately all the details of a meteor seen only in the observer's peripheral vision.

Constellation in which seen

The BAA Meteor Section suggests that, rather than recording reliability, observers should indicate in which part of the sky the meteor was seen. This can be useful in identifying potentially photographed meteors when the negatives are later scanned (see Chapter 6), or in tying together reports of particular bright meteors which may have been seen simultaneously by several observers.

Other useful details

While the meteor characteristics described above may be regarded as absolutely fundamental, and should be recorded for each meteor, there are other details which the observer might find it interesting to add.

Many observers record *colours* of meteors. I frequently see bright meteors as yellowish; other observers often note a greenish tinge, though pure green meteors are uncommon. Fainter meteors are usually perceived as white in colour. Presumably faint meteors have less effect on the colour receptors (the cones) in the human eye, and do not elicit a colour response. Red is occasionally seen, most often among slow meteors. Taurid meteors sometimes show blue and red colours, while I have also seen purple "sparks" in Perseids.

Other interesting characteristics to note are *flares* in brightness along the path of the meteor. These again reveal something about the structure of the incoming particle. Perseid meteors are noted for the terminal flares which mark their extinction, while other showers, such as the Taurids, produce meteors which may flicker. Unusual meteors, such as those which may appear curved, or point-source (head-on) events, may also be noted.

Naked-eye visual meteor watches, then, are quite straightforward, and can yield results of some scientific value. I must stress again, however, that the observer's main reason for carrying out a meteor watch should always be the enjoyment derived from doing so – meteor observing is not a chore! Whether it is your intention simply to observe only on those nights around the maxima of major showers, or whether you wish to add to the limited amount of data available for the quieter parts of the year, this principle applies. Remember too that if you wish your results to have some scientific as well as personal value, they should be communicated to one of the organizations which collect and coordinate such work. Figure 4.4 is a sample BAA Meteor Section Report Form on which are entered the results from an actual watch.

OFFICE USE ONLY	
REC'D	ACK'D

Date 11-12/8/91   Observer(s): NEIL BONE

Observing Site: APULDRAM   50° 49'·8 N   0° 48'·3 W   Sheet 1 of 1

Correspondence Address: THE HAREPATH, MILE END LANE, APULDRAM, CHICHESTER, WEST SUSSEX, PO20 7DZ

Observing Conditions: PATCHY CIRRUS 10%   Stellar Lim. Mag.: 5·0

Watch Times: Start. 2150 UT   End. 2250 UT   Duration. 1H 00M

Code №	Time U.T.	Magni-tude	Name Shower or if Sporadic	Constellation(s) in which seen	Train Details & time to fade (secs.)	Notes
1	2200	0	PERSEID	PEG-AQR	1 SEC	YELLOW - RED MEDIUM SPEED LONG
2	2205	3	SPORADIC	AQR		YELLOW FAST
3	2209	1	PERSEID	LYR- HER	WAKE	WHITE FAST
4	2211	3	PERSEID	PEG		WHITE FAST
5	2216	2	α CAPRICORNID	SCT		WHITE SLOW
6	2218	1	PERSEID	CYG	2 SEC	YELLOW FAST
7	2220	3	PERSEID	CYG		WHITE FAST
8	2230	0	PERSEID	LAC	2 SEC	YELLOW MEDIUM SPEED
9	2230	2	S AQUARID - S	PEG		YELLOW SLOW
10	2230	3	SPORADIC	AQL		WHITE
11	2236	2	PERSEID	AQR		WHITE FAST
12	2239	2	PERSEID	OPH		YELLOW FAST
13	2241	4	SPORADIC	CEP		WHITE FAST
14	2241	2	PERSEID	AQL		WHITE FAST
15	2245	3	SPORADIC	CYG		WHITE SHORT
16	2247	3	PERSEID	CAP		WHITE FAST
17	2250	1	PERSEID	AQL		YELLOW FAST

ALL WITHIN A FEW SECONDS (bracket spanning rows 8, 9, 10)

Use back of form for details of plotted paths of any meteors listed above

Figure 4.4 *An example of a completed meteor report form, containing details of an actual watch performed by the author during the Perseid shower in 1991.*

Group meteor observing

The field of view of a single observer probably covers about a fifth of the sky, and meteors appearing outside it will obviously be missed. One solution to this limitation is for observers to work together in groups from a single location, with each covering a previously agreed patch of the sky. Depending on how many observers can be recruited for a night's watch, complete coverage of the whole sky at the optimal 50° elevation can sometimes be achieved. Local astronomical societies sometimes organize whole-sky group watches at the maxima of major showers such as the Perseids or Geminids. As well as providing an opportunity to make useful observations, such group events have a worthwhile social angle and can become a focus for activities within a society.

Basic requirements for each observer in a group watch are exactly the same as for an individual observer, and the method of watching the sky and recording the details of any meteors seen is the same too.

Six observers seems to be the maximum desirable size for a group. If more are available, it is worth setting up two separate groups, whose results it may be interesting to compare later. Observers are usually arranged as in Figure 4.5, so that there is a little overlap between fields of view. One member of the group, the *central recorder*, who is not actively observing, takes responsibility for recording all meteors seen by the observers. This procedure cuts down dead time. On seeing a meteor, an observer should call for the recorder's attention, and give the details, in order, quickly and clearly. At busy times a queue can build up. The recorder is responsible for logging the time of each sighting, and should also note the identity (by initials) of the observer or observers who made the sighting. The central recorder should be changed each hour to allow everyone in the group the chance to do some observing.

One cautionary note: observers' perceptions vary, and when two or more witness the same meteor, they will not necessarily agree on the

Figure 4.5 *The ideal arrangement for six observers in a group meteor watch.*

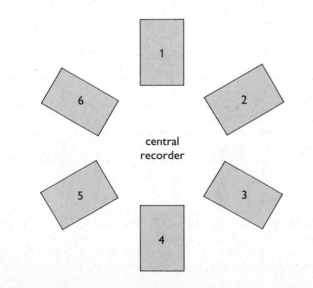

details. Avoid haggling, which in my experience has inevitably resulted in meteors being ascribed much brighter magnitudes than they really had. Inexperienced observers in a group, particularly, may be carried away by the sight of a fairly bright meteor, magnitude −2 say, and insist on it being logged as magnitude −5 or brighter. The safest approach seems to be for each observer's account, with his or her initials, to be recorded separately.

Fireballs

Most meteor watches yield a reasonable crop of moderately bright events, around magnitudes 0 to +2. Negative-magnitude meteors (comparable in brightness to, say, Jupiter) are less common, except on nights when major showers such as the Perseids or Geminids are active. Occasionally, however, the regular observer may see much brighter events – fireballs.

A fireball is, by definition, a meteor brighter than the planet Venus – in other words, of magnitude −5 or brighter. The Harvard photographic surveys (see Chapter 3) suggested that such events represent only a very small percentage of all meteors. Fireballs tend to be more common during the major annual showers; in particular, large numbers of bright objects are always reported during the Perseids and Geminids.

Perhaps of greater interest are sporadic fireballs, which can occur at any time of year. Some of these events may be associated with the arrival of bodies sufficiently large to survive atmospheric passage and land as meteorites. Significantly, and presumably reflecting both the small size and friable nature of their particles, none of the major showers has ever been found to produce a meteorite-dropping event.

The scientific importance of new meteorite finds in turn makes it important that fireballs are accurately observed. Reconstruction of the body's atmospheric trajectory may, rarely, allow any resulting debris to be collected.

Patrol photography has been used in the past for such work, but, as mentioned in Chapter 3, it has been largely neglected since the 1970s. Consequently most fireball reports are chance sightings made by visual observers, amateurs engaged in other forms of astronomical observation, or, most frequently, members of the public. Regular meteor observers should, naturally, be aware of the worthwhile details to be recorded when a fireball is seen:

Time of appearance The time at which a fireball is seen should be recorded as for any other meteor. Accurate timings help in positive identification of the fireball.

Magnitude Fireball magnitudes are, naturally, very difficult to estimate, since there are few comparison objects in the appropriate range. Brightness could be quoted in terms of the half Moon (magnitude about −8) or full Moon (−12). It may also be useful to note whether a fireball cast shadows, or lit up the ground.

Light variations Fireballs often show flares along their path, which again may be useful in confirming the object's identity in multiple records. Any strong colours should also be noted.

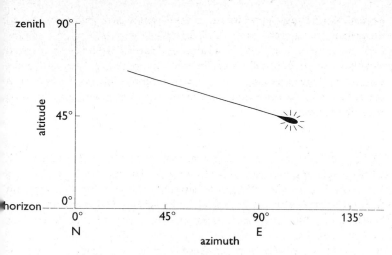

Figure 4.6 *Altitude and azimuth for a fireball.*

Apparent trajectory Probably the most important fireball observations are those which give positional details. Experienced observers may be able to provide a fireball's start and end positions against the background stars. Equally useful are details of the altitude and azimuth of the start and end points, as illustrated in Figure 4.6.

Altitude is given in degrees above the horizon, and can be estimated, from memory, by using a 15 cm (6 inch) ruler. Held at arm's length, the straight-edge subtends an angle of 20° against the sky background. An alternative rule-of-thumb is that a fully outstretched hand at arm's length also subtends 20°. Azimuth is measured from 000° at due north through 090° at due east, and on to 360° back at due north again. Polaris is, of course, a convenient indicator of the zero point of azimuth.

Visual positional estimates by experienced observers may locate a fireball's path against the sky to within about 0.5°, but lower accuracy is more usual – particularly for rare daytime events. Such events can have their positions fixed with reference to chimneys, television aerials, or other local features, whose positions can later be translated into altazimuth coordinates.

Sound Some fireballs which remain luminous to altitudes as low as 30 km (20 miles) in the stratosphere may be followed, after a few minutes' delay, by rumbles like distant thunder, crackling noises, or sonic booms. Technically, a fireball which is accompanied by sound effects – which should, of course, also be noted in reports – is referred to as a *bolide*.

Reporting fireballs

Bright fireballs should be reported as soon as possible to a coordinating body such as the American Meteor Society or the Meteor Section of the BAA, so that follow-up investigations can be started quickly. This is particularly important if members of the public are to be interviewed for their eyewitness accounts – it is essential to collect such accounts while the event is still reasonably fresh in peoples' minds.

Collecting reports from the public Many reports come not from amateur astronomers, but from members of the public – especially people taking their dog for an evening walk! Useful details can sometimes be obtained from such witnesses, and astronomical societies and clubs can play a useful role by placing appeals for information in their local newspaper. Newspapers sometimes carry eyewitness reports of fireballs seen by excited members of the public, and may welcome the opportunity to extend the news story.

Casual observers should be encouraged to recount their sighting to an interviewer who should attempt to elicit the details an experienced observer would record. In particular, if some idea of observer location, time, and rough altazimuth coordinates for the fireball can be obtained from several witnesses, the effort may prove particularly worthwhile, yielding an approximate ground track for the object.

An interesting subjective effect often found in reports from inexperienced observers is that the fireball appeared to come down nearby. In most cases, however, fireballs are seen from some considerable distance.

Unusual meteors

The various coordinating bodies occasionally receive reports of unusual meteors. Meteors appearing in pairs along parallel paths are likely to be the result of the virtually simultaneous arrival of two very recently separated fragments of a single larger meteoroid. Reports of curved meteors are best explained as illusions of perception. Many apparently curved meteors are spotted in the observer's peripheral vision. As the observer turns rapidly towards the direction of the meteor, the momentary disorientation may be sufficient to produce the illusion of a curved trajectory.

More enigmatic are those meteors which allegedly produce sound simultaneously with their appearance (as distinct from the delayed clicks and rumbles from bolides). Such observations are somewhat controversial, but the relatively few people who have made them usually remain adamant that they heard hissing or similar sounds which they associated with the meteor, at the time of the meteor's appearance. The obvious problem with this is that sound travels much more slowly than light, so any sonic effects should follow the meteor's appearance by several minutes. Very little of a meteor's energy, in any case, should be released in the form of sound, and by the time this has been attenuated by the atmosphere between meteor and observer, it should be barely audible, if at all. None the less, observers do, rarely, report hearing such sound – usually from remote, otherwise quiet locations.

One theory is that the energy is propagated electromagnetically, being amplified and "re-transmitted" by suitable nearby sources, such as metal fence wires. This possibility of *electrophonic* meteor sound has many adherents. Others, myself included, are more sceptical about the possibility of meteor sound in the first place. Perhaps years of watching firework rockets have somehow conditioned us to expect sound from rapidly moving lights in the sky. Perhaps, also, the "quiet" locations from which such observations are made are less tranquil than the observer believes

A final curiosity is the occasional reporting of "black meteors." These seem to occur most commonly just after a watch has started, or towards the end of a long stint. Almost certainly, the "black meteor" illusion is a physiological effect in the eye, either as it settles down to dark adaption, or as fatigue sets in later in the watch.

In the next chapter I look at the major annual meteor showers and their observational characteristics. Most of these details are the result of careful, standardized processing of naked-eye meteor reports. While reports should certainly be submitted to a national coordinating body such as the BAA Meteor Section or American Meteor Society for inclusion in large-scale analyses, there is no harm in the observer attempting a preliminary analysis of his or her data, provided the limitations of using a single set of results are recognized. It would be foolhardy to draw profound conclusions from a single observer's work, and it is for this reason that the major organizations prefer to collect as many sets of data as possible before publishing reports on a given shower's performance at a particular return.

What it all means: analysing visual meteor data

As mentioned previously, there is little point in trying to conclude too much from watches made under poor sky conditions. As we shall see, skies with limiting magnitude worse than 5.0 lead to artificially high calculated rates from low observed meteor numbers, and watches made under such conditions should not be used in analyses. Similarly, average cloud cover in excess of 20 percent has an adverse effect.

The first and most basic analysis one can carry out is of estimated meteor *magnitudes*. As mentioned earlier, an experienced observer can reliably give these to the nearest whole magnitude. We can look at magnitudes in a couple of ways. Most obviously, a simple averaging will usually reveal that shower meteors tend to be slightly brighter on average than the *contemporaneous sporadic background* (the sporadic meteors visible at the same time of the year). For example, results obtained by a team of BAA observers in Scotland and the north of England in 1983 gave a mean Perseid magnitude of 1.51, compared with a sporadic mean of 2.84. Geminid meteors observed in 1990, during another BAA project, were similarly a little brighter on average: the mean magnitude for Geminids was 2.04, while sporadics observed during the same watches had a mean magnitude of 2.35.

More informative is an examination of the *proportion* of meteors in each magnitude class. Histograms of these can easily be prepared. As an example, histograms of the 1983 Perseids and of sporadics from the same period (Figure 4.7) show the shower to be slightly depleted in the fainter range relative to the random background. Over time, a significant proportion of the smaller meteoroids, which give rise to fainter meteors in the atmosphere, have been lost from the Perseid stream through the influence of the Poynting–Robertson effect, described in Chapter 1.

Obviously, for a magnitude mean or distribution to have any statistical worth, it should be based on as many meteors as possible. It is not,

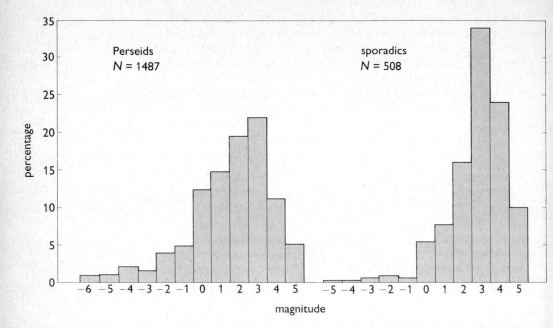

Figure 4.7 *Comparison of sporadic and Perseid magnitude distributions from 1983 observations. The Perseids are slightly enriched, in proportion, in brighter meteors relative to the sporadic background; but the proportion of faint Perseids is lower.*

for example, sound practice to base a magnitude distribution on only ten observed meteors. To show the validity of your observations, if presented in this form in a club magazine or similar publication, you should always quote the value of N – the number of meteors on which the graph is based – as in Figure 4.7.

The magnitude distribution is important with respect to the *population index* (r), used later in calculating corrected hourly rates for shower meteors. Loosely, the population index describes the ratio between meteors in adjacent magnitude classes (populations). As a general rule, smaller meteoroids – that is, those which produce fainter meteors – should be more common than larger meteoroids. During watches, however, this rule appears to break down somewhat since the observers tend to miss many of the faintest meteors. In practice it is more convenient to determine the population index from the average magnitude, using a graphical relationship derived by the Czechoslovakian meteor astronomer M. Krešakova.

A high r value is typical of older meteor showers, where the faint end of the magnitude distribution is depleted. Younger streams have smaller r values. Typical r values for the major annual showers are presented in the next chapter, in Table 5.1 (see pp. 86–7). Research by George Spalding suggests that most major showers have r close to 2.25.

The magnitude distribution, then, tells us something about the particle *size* distribution in meteor streams. Other observational characteristics can tell us something about the nature of these particles. Meteoroids from cometary sources tend to be of low density, and fragment very rapidly on impact with the upper atmosphere. *Persistent train* phenomena are to some extent associated with this process. Interestingly, rather few Geminid meteors, which are produced by dense particles of probably a more "rocky" nature, show persistent trains.

Persistent trains also reflect the geocentric velocity of the incoming meteoroid (that is, its velocity relative to the Earth). Meteoroids impacting at high velocities impart more energy to the atmosphere. For example, the particle size distributions of Perseid and Taurid meteors are broadly similar. Perseids impact at a velocity of 60 km/s (37 mile/s), Taurids at 35 km/s (22 mile/s); usually about 30 percent of Perseids leave persistent trains, while only 10 percent of Taurids do so. Orionids and Leonids have very high geocentric velocities – their retrograde orbits produce almost the highest collision velocities possible between meteoroids and the upper atmosphere – and similarly have a high proportion of meteors leaving persistent trains.

These clues to the nature of the particles apart, visual observations are also of considerable value in establishing the spatial density of meteoroids within a stream. These data are revealed through the corrected *zenithal hourly rate* (ZHR). Activity in meteor showers can often show variations over timescales of hours, and hourly counts are therefore an appropriate means of assessment.

The ZHR makes allowance for a number of observational factors which are beyond the control of the meteor watcher. By definition, the ZHR is the theoretical number of meteors which would be seen by a *single* alert observer, watching under perfectly cloudless skies, in the absence of haze (the "standard sky" has a limiting magnitude of 6.5), and with the shower radiant overhead. In practice, of course, this ideal set of conditions seldom, if ever, comes together. Haze, for instance, frequently intrudes, while watches can be troubled from time to time by patchy cloud. As the Earth turns on its axis, the observer will find the radiant presented at differing altitudes.

Radiant altitude is an important factor. During watches in August, for example, observations will often commence late in the evening, around 22.00 hours local time. At this time the Perseid radiant is low in the northeast, and relatively few Perseids will be observed. As the night goes on, and the radiant rises higher into the sky, more and more Perseids are usually seen each hour until dawn. In simple terms, when the radiant is low a significant number of shower meteors are out of sight beyond the observer's horizon, or lost in the haze near the horizon, as illustrated in Figure 4.8. This problem becomes less pronounced as the radiant climbs. To a reasonable approximation, the influence of the radiant's altitude on observed shower activity is given by $1/\sin a$, where a is the radiant's altitude in degrees.

The altitude a can be worked out by a little fairly straightforward trigonometry. In Chapter 5, tables of radiant altitudes at intervals through the night are given for the major showers, as an aid to ZHR calculation. For those seeking more precise values, I can recommend the routines presented in Peter Duffet-Smith's book *Practical Astronomy*

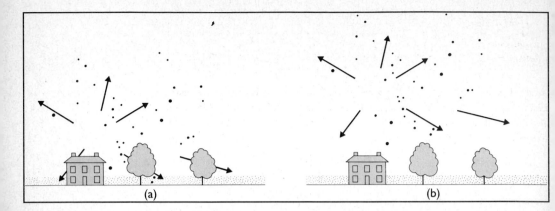

Figure 4.8 *Observed shower meteor rates generally increase as the radiant rises higher in the sky. In early evening (a), for example, the Perseid radiant is low, and many meteors are obscured behind obstructions on the horizon, or are lost in the haze. Later on (b), when the radiant is higher, more meteors are visible.*

with your Calculator. Modern scientific calculators have all the necessary trigonometric function keys for these calculations.

The other principal determinant of observed shower activity is the sky limiting magnitude. This is where r, the population index factor, comes in. Correction of observed rates to ZHR, allowing for the limiting magnitude (LM), is given by $r^{6.5-\text{LM}}$ (again, a computation readily performed on modern scientific calculators). Obviously, the closer the limiting magnitude is to 6.5, the better. Observations under poor, hazy skies of limiting magnitude worse than 5.0 will produce spuriously high ZHRs from low observed meteor numbers.

Cloud has an obvious influence on observed rates. Basically, it is not really worth trying to calculate ZHR from observations made in conditions where the cloud average in the hourly interval is greater than 20 percent. Cloud correction is given by $1/(1-x)$, where x is the cloud cover expressed as a decimal, averaged over the hour.

It is usual to ascribe a statistical error limit to quoted ZHRs, given by ZHR/$\sqrt{N}$, where N is the observed number of meteors on which the ZHR is based. Obviously, the more meteors observed, the less uncertain the ZHR. If very large numbers of observations can be combined, statistical errors are kept to a minimum.

Putting all this together, we might, for example, have a Perseid watch as follows (calculations are rounded appropriately):

Date: 1983 August 11–12   Observer; Neil Bone
Location: Campbeltown, Scotland; 55°25′N, 5°36′W

UT	LM	Sporadics	Alpha Cygnids	Perseids
2155–2255	5.8	4	2	33
2255–2355	6.0	5	1	36
2355–0055	6.0	7	0	36

(5 percent cloud in last hour)

Table 4.3 *Correction factors for group watches.*								
Number of observers	1	2	3	4	5	6	7	8
Factor	1.00	0.56	0.41	0.36	0.32	0.28	0.25	0.22

The Perseid ZHR for the first hour is found as follows. The mean radiant altitude is 36.6°, giving a correction factor of $1/\sin 36.6° = 1.68$. A factor of $r^{6.5-LM} = 2.35^{0.7} = 1.82$ takes account of the limiting magnitude. Therefore, $ZHR = 1.68 \times 1.82 \times 33 = 100.9$. The error, given by $ZHR/\sqrt{N}$, is ± 17.6, so the result for the first hour is

$$ZHR = 100.9 \pm 17.6$$

For the second hour,

$$ZHR = 80.5 \pm 13.4$$

For the third hour we have to take account of cloud in addition to the other factors: $1/(1 - x) = 1/(1 - 0.05) = 1.05$, $1/\sin 48.7° = 1.33$, and $2.35^{0.5} = 1.53$; so

$$ZHR = 1.05 \times 1.33 \times 1.53 \times 36 = 76.9 \pm 12.8$$

Sporadic rates are also corrected, to CHR (corrected hourly rate), by making allowance for limiting magnitude and cloud. A value of $r = 3.42$ is reasonable for sporadics. Thus, in the examples above the CHRs are respectively 9.4 ± 4.7, 9.3 ± 4.2, and 13.6 ± 5.1. More sophisticated analyses take account of the changing sporadic rate through the night, and of the deviations of observed rates from the *expected* diurnal rate curve (see for example Figure 2.2, p. 38)

It is normal to use single-observer counts to derive a ZHR, but it can, with some limitations, be obtained from group watches using the correction factors in Table 4.3.

In the next chapter I look at the principal showers of the year. Much of what we know about these showers, their activity profiles, and their particle composition is based on the kind of analysis described above. Often, when a particular shower is favourable with respect to moonlight, the national organizations which collect meteor data will mount special observing campaigns. As outlined above, the benefits of having as many results as possible from which to draw conclusions are obvious!

CHAPTER 5

The meteor observer's year

As discussed in Chapter 2, there are a number of reasons why observed meteor activity varies over the course of the year. The main reason is that the Earth encounters more concentrated streams of (usually) cometary debris at certain points in its orbit around the Sun. At such times, the annual meteor showers are seen. Observers will find that their watches are most productive during the more active showers, and particularly around those showers' peak nights. Watches by experienced observers at other times, however, are not to be despised, as they can tell us much about the zodiacal dust cloud which fills the inner Solar System.

Having covered the essentials of simple naked-eye observing in Chapter 4, we now proceed to a review of what is on offer meteorically as the year unfolds. Also, we shall find that some showers are active only in certain years; such showers may even, at times, exceed the regular annual displays in activity.

Moonlight Apart from the weather, how well a shower can be observed in a particular year is determined mainly by the Moon. A bright Moon will swamp the fainter meteors, making watches rather unproductive and diminishing the reliability of any eventual ZHR analyses. To some degree, the effect of moonlight depends on the observer's latitude and the time of year.

The Moon follows more or less the same ecliptic path around the sky as do the Sun and the major planets. With respect to observing conditions, the Moon's position is quite significant at higher northern latitudes such as those of the British Isles and Canada. Here, a high-riding spring crescent Moon can seriously hamper attempts to observe, say, the Lyrids. The ecliptic in the western sky after sunset at this time of year cuts a steep angle with the horizon, such that even the crescent Moon, not far in elongation from the Sun in the sky, can remain above the horizon until well after midnight. Conversely, the waning gibbous Moon a few days after full Moon may not rise until quite late, allowing some observations to be made at times when moonlight might otherwise have been expected to be a problem.

The situation is reversed in the autumn, when the ecliptic in the eastern sky makes a shallow angle with the horizon, so that the waning gibbous Moon may rise early, a little further north each evening, for several successive nights, giving serious problems for Perseid watchers.

Observers at tropical latitudes have more or less the same moonlight conditions each lunation, with less marked seasonal variations.

Table 5.1 lists the main showers, and should be consulted before setting out to observe, to check which are active. Other details of a shower, including the moonlight conditions, for a particular year may be found in the annual publications mentioned on p. 66. Where quoted in this chapter, the solar longitude (see Chapter 1) is for the epoch of 2000.0, adopted as a standard for Solar System positions after November 1991. Zenithal hourly rates may vary from year to year, and the values given here are based on typical returns observed in the 1980s.

Quadrantids (January 1–6)

The new year

The opening days of the year bring to northern hemisphere observers one of the three most active regular showers. Unfortunately, poor weather conditions at this time often result in the shower, active principally around a very short peak, being completely missed in some years. The sharp maximum occurs around a solar longitude of 283.4°, usually on January 3–4.

A meteor shower is usually named for the constellation in which its radiant lies, a convention introduced by Schiaparelli in the 1860s. If more than one radiant is active from a constellation at a given time (and this can get complicated, as we shall see for July), each shower is named for the prominent star closest to the radiant position at maximum. The Quadrantids are something of an exception. Their radiant lies in the now-defunct constellation of Quadrans Muralis, which along with a few others was removed from the over-cluttered heavens in 1922 when the International Astronomical Union adopted the modern list of 88 offici- ally recognized constellations. The region once occupied by Quadrans Muralis lies at the junction between Hercules, Boötes, and Ursa Major.

The radiant at maximum is a few degrees northwest of the triangle of stars formed by Tau, Phi, and Nu Herculis (Figure 5.1). For observers at latitudes above 40°, this radiant is circumpolar. If the peak is at a time when the radiant is low in the northern sky, however, observed rates can be relatively poor. Best rates are obtained in the early morning hours, as the radiant climbs high in the northeast, and it is often worth waiting until 02.00 hours local time before starting a serious watch for the Quadrantids. Unfortunately, the limited span of high Quadrantid activity does not always allow this luxury. Table 5.2 gives the Quadrantid radiant altitude as a function of local time from various latitudes.

Observations in the 1970s, analysed by the BAA's then Meteor Section Director Keith Hindley, suggested a peak ZHR of around 80–100, with a sharp rise and decay. The time taken for activity to decline to half its peak value was found to be about 4 hours. During the 1980s, good Quadrantid activity was recorded on a number of occasions. The 1981 and 1984 returns were well covered by BAA observers. Observed rates in 1984 were excellent in the pre-dawn hours: a group of six observers in Wales recorded 52 Quadrantids in a 58-minute watch just before dawn. I was fortunate to see the shower rising towards peak in excellent conditions on 1985 January 2–3, with observed rates in the hour before dawn of 29 meteors per hour, corresponding to a corrected ZHR of 90.

Table 5.1 *The principal annual nighttime meteor showers.*

Shower	Activity limits	Approximate maximum Date	Solar longitude	ZHR	Radiant position Peak at maximum RA	dec	Daily drift RA	dec
Quadrantids	Jan 1–6	Jan 3	283.4°	100?	15h 28m	+50°		
Alpha Aurigids	Jan–Feb	Feb 6–9		10	4h 56m	+43°		
Virginids	Mar–Apr	Apr 12	022°	5	14h 04m	−09°		
				5	13h 36m	−11°		
Lyrids	Apr 19–25	Apr 21	032.1°	10–15	18h 08m	+32°	+1.1°	0.0°
Eta Aquarids	Apr 24 – May 20	May 5	045°	35	22h 20m	−01°	+0.9°	+0.4°
Alpha Scorpiids	Apr 20 – May 19	Apr 28	038°	5	16h 32m	−24°	+0.5°	−0.2°
		May 13	052°		16h 04m	−24°		
Ophiuchids	May–Jun	Jun 9	079°	5	17h 56m	−23°		
		Jun 19	089°		17h 20m	−20°		
Alpha Cygnids	Jun–Aug	Jul 21	118°	5	21h 00m	+48°		
		Aug 21	148°					
Capricornids	Jul 5 – Aug 20	Jul 8	106°	5	20h 44m	−15°		
		Jul 15	113°		21h 00m	−15°		
		Jul 26	123°					
Alpha Capricornids	Jul 15 – Aug 20	Aug 2	130°	5	20h 36m	−10°	+0.9°	+0.3°
Delta Aquarids	Jul 15 – Aug 20	Jul 29	126°	20–25	22h 36m	−17°	+0.8°	+0.2°
		Aug 6	134°	10	23h 04m	+02°	+1.0°	+0.2°
Iota Aquarids	Jul–Aug	Aug 6	134°	10	22h 10m	−15°	+1.1°	+0.2°
					22h 04m	−06°	+1.0°	+0.1°
Piscis Australids	Jul–Aug	Jul 31	128°	5	22h 40m	−30°		
Perseids	Jul 25 – Aug 20	Aug 12	140.0°	80	3h 04m	+58°	+1.4°	+0.1°
Alpha Aurigids	Aug–Oct	Aug 28	156°	10	4h 56m	+43°		
		Sep 15	174°		4h 52m	+41°		
Piscids	Sep–Oct	Sep 8	166°	10?	0h 36m	+07°		
		Sep 21	178°		0h 24m	0°		
		Oct 13	200°		1h 44m	+14°		
Orionids	Oct 15 – Nov 2	Oct 21	209°	30	6h 24m	+15°	+1.2°	+0.1°
Taurids	Oct 15 – Nov 25	Nov 3	221°	10	3h 44m	+14°	+0.8°	+0.2°
					3h 44m	+22°	+0.8°	+0.1°
Leonids	Nov 15–20	Nov 17	235.4°	Variable	10h 08m	+22°	+0.7°	−0.4°
Geminids	Dec 7–15	Dec 13	262.0°	100?	7h 28m	+32°	+1.1°	−0.1°
Ursids	Dec 19–24	Dec 23	271°	5–10?	14h 28m	+78°	+0.9°	−0.5°

*r is the population index (see p. 82): higher r means more brighter and fewer fainter meteors.

Clear skies over virtually the whole of the British Isles on 1986 January 3–4 allowed a great amount of Quadrantid data to be collected. Sadly, rates seem to have been at their best during the afternoon daylight. By nighttime the shower was on the decline, and while good conditions resulted in reasonable observed rates, activity was never spectacular. In the early evening, Quadrantid rates of about 15–20 meteors per hour were seen. Many UK observers missed this early activity since they were, understandably, concentrating on observations of Comet P/Halley, which was low in the southwestern sky, and about to be lost from view at such latitudes. By the early morning hours, when the radiant was beginning to climb higher into the northeastern sky, rates had declined to about 10 meteors per hour.

As highlighted by the return of 1986, observations throughout the 1980s revealed that the peak naked-eye Quadrantid activity was occurring

Characteristics*

Strong activity around brief maximum. Bright meteors are bluish or yellow–green. $r = 2.1$
Slow, bright
Slow, sometimes bright. Complex radiant

Fast. Occasional strong outbursts, as in 1922 and 1982. $r = 2.9$
Fast, often with persistent trains. Difficult for observers at high northern latitudes. $r = 2.3$
Weak shower, part of complex group of radiants producing activity from the region of the ecliptic throughout summer

Weak ecliptic stream

Apparently stationary, weak radiant producing steady activity throughout late summer

Weak, possibly with several sub-maxima. Slow, bright, yellowish meteors

Another weak southern radiant with sub-maxima. Several long, bright meteors and fireballs each year
Double radiant, the southern one the more active. Good observed rates at more southerly latitudes

Double radiant. Weak with mainly fast, faint meteors

Poorly observed southern shower
Very rich shower, producing many bright events with persistent trains. High rates attract many observers around maximum.
    Excellent target for photographers. $r = 2.35$
Low activity, occasional fireballs

Sustained, low activity from two radiants near the ecliptic

Very fast. Brighter Orionids often leave persistent trains. Broad peak with several sub-maxima over a few days. $r = 2.25$
Double radiant. Steady, low rates; broad maximum. Slow, occasionally bright meteors. $r = 2.25$

Periodic storms, and good rates generally, around perihelion of parent comet, P/Tempel–Tuttle (1966, 1999?). Very fast
    meteors, some fine trains. $r = 2.5$
Perhaps the best shower in most "normal" years. Fairly fast meteors, sometimes of long duration. Good target for
    photographers. Bright meteors are yellowish. $r = 2.44$
Poorly observed. Occasionally produces good rates, as in 1945 and 1986

somewhat earlier than expected. The Quadrantids are a good example of a stream whose orbit is rapidly evolving. Close approaches of the stream to Jupiter near its aphelion have resulted in some spread of its meteoroids. The orbit appears to oscillate up and down relative to the plane of the ecliptic, such that it was formerly encountered by the Earth (as an Aquarid radiant in July) in the early centuries AD. The stream was then "missed" for over a thousand years, before swinging back down to be encountered at its descending node by the Earth from the 1700s onwards. We shall "lose" the Quadrantids again in another three or four hundred years.

These relatively rapid variations in the stream's orbit frustrate attempts to determine the parent comet. The meteors do show several characteristics that point to a cometary origin: for example, around 11 percent of Quadrantids leave persistent trains. The meteors are of

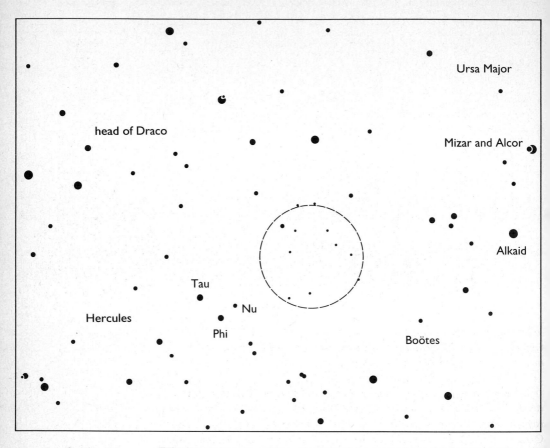

Figure 5.1 *Quadrantid radiant position. (On this and other radiant charts in this chapter, stars are shown down to about fourth magnitude. Fainter naked-eye stars are shown near the radiant. Variable stars are shown at their maximum brightness.)*

Table 5.2 *Quadrantid radiant altitudes (in degrees), for various latitudes.*

Local time (h)	Latitude							
	25°N	30°N	35°N	40°N	45°N	50°N	55°N	60°N
17	—	—	—	—	—	—	24.1	28.2
18	—	—	1.5	6.1	10.7	15.2	19.8	24.3
19	—	—	—	2.3	7.1	12.0	16.8	21.6
20	—	—	—	0.3	5.2	10.3	15.2	20.2
21	—	—	—	0.2	5.2	10.2	15.2	20.1
22	—	—	—	2.0	6.9	11.7	16.6	21.4
23	—	—	1.0	5.6	10.2	14.8	19.4	24.0
00	0	2.2	6.5	10.8	15.1	19.4	23.6	27.7
01	5.5	9.5	13.5	17.4	21.3	25.1	28.9	32.5
02	13.8	17.5	21.2	24.8	28.3	31.7	35.0	38.0
03	23.2	26.8	30.2	33.5	36.6	39.6	42.2	44.6
04	32.8	36.3	39.5	42.6	45.4	47.9	50.0	51.6
05	42.2	45.9	49.2	52.1	54.6	56.7	58.2	60.0
06	—	—	58.7	61.8	64.1	65.9	66.7	66.5
07	—	—	—	—	—	75.2	75.3	73.6

moderate speed, having a geocentric velocity of 41 km/s (25 mile/s), and are often white or blue-white in colour. Relative to the contemporaneous sporadic background, Quadrantids are only slightly brighter, on average;

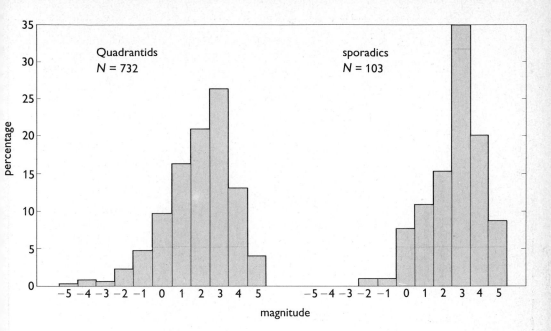

Figure 5.2 *A comparison of magnitude distributions for Quadrantid and sporadic meteors observed at the shower's excellent return in 1992.*

the precise magnitude distribution obtained during a Quadrantid watch depends heavily on solar longitude. Hindley's work in the 1970s revealed some evidence of mass-sorting, such that rates for the fainter Quadrantids peaked earliest, followed by those in the naked-eye range, and finally bright meteors in the photographic range.

Observers in northwest Europe were treated to an exceptionally fine Quadrantid return in the early morning hours of 1992 January 4 – again suggesting that the peak is getting earlier. Corrected ZHRs touched 130 around 0300 UT (solar longitude 283.1°), and observers enjoying the best conditions recorded Quadrantids at rates of 2–3 per minute at times. Several bright events were also seen, including a number of yellow–green fireballs. Figure 5.2 shows the magnitude distributions for Quadrantid and sporadic meteors seen during watches by BAA observers in 1992.

January–March minor showers

Once the Quadrantids are past, the year sinks into one of its low periods for meteor observers, with no major shower activity until late in the spring. A couple of minor showers, whose activity is barely distinguishable from the sporadic background, are present but their low activity makes them challenging targets, best left to the truly dedicated.

It is at this time of year, particularly, that the often-quoted "average" rate of 6–7 meteors per hour comes into question. More typically, an hour of meteor watching on a February or early March night will

The low months: January to early March

reward the observer with only two or three meteors. Since few observations are made at this time of year, however, watches by the dedicated, experienced few can be of considerable value in maintaining coverage of the overall annual activity pattern.

The *Alpha Aurigids* are a minor shower with a reasonable span of activity from late January to mid-February. Apparently rich in reasonably bright meteors, it seems to be an aged and rather depleted shower, producing only one or two fairly slow meteors each hour.

Low-activity showers such as this raise the question: when is a shower not a shower? Work by David Hughes at Sheffield University suggests that there may be many weak streams producing only one meteor every couple of hours. Such levels of activity are all but lost in the contemporaneous sporadic background, although perhaps as many as a quarter of all "sporadic" meteors belong to these very minor streams. For most practical purposes, it is impossible to distinguish a shower producing ZHRs lower than about 3 meteors per hour from the background.

Perhaps a case in point is the fireball streams which are sometimes proposed to exist. Despite the generally low levels of meteor activity in the spring months, this seems to be one of the best times of year for bright events: the period February to April may be a "fireball season." The idea has been put forward that the spring fireballs might be members of one of the very minor streams Hughes has proposed. Much more work is required before this can be confirmed. Systematic photographic observations, for example, would help to determine whether there is a common radiant for these events, while more advanced techniques could be used to calculate orbits for spring fireballs.

The spring

By March, there are signs of activity from the various *Virginid* radiants. This is a complex set of streams – or perhaps a single ancient stream, split long ago into several components by gravitational perturbations by the planets – with orbits lying close to the plane of the ecliptic. Activity is seen, at low levels, from late February through to mid-April. There are two main radiants: one near Spica, the other not far from Kappa Virginis. Many observers have reported another radiant in the Virgo Bowl, near Gamma Virginis, active during early April. From the British Isles, observations are again favoured after midnight, as only then do the radiants attain a reasonable altitude above the horizon. US observers, particularly in the southern states, are rather better placed.

Typical rates are low – in late March and early April, the observer may record only one or two Virginids per hour. Meteors from this shower are often slow and long, the brighter examples showing flares along their paths. Overall, the Virginid magnitude distribution is quite similar to that of the contemporaneous sporadic background. While Virginid activity is generally low, the shower can provide the occasional surprise. During a watch on 1979 March 28–29, I was lucky enough to record four negative-magnitude Virginids in a 22-minute burst. Later that night another very bright Virginid, of magnitude −4, flared across my northeastern sky, leaving behind it a persistent train.

Lyrids (April 19–25)

After the long wait since January, April's Lyrids come as a welcome change from the generally low early spring activity. These meteors, emanating from a point about 10° southwest of the brilliant star Vega (Figure 5.3), the brightest member of the Summer Triangle, are usually fast (geocentric velocity 49 km/s, 30 mile/s), and a reasonable proportion (around 8 percent) leave persistent trains. Magnitude distributions calculated from observations made during the 1980s suggest that the Lyrids are fairly rich in bright meteors relative to the contemporaneous sporadic background.

The peak ZHR is typically 10–15 (Figure 5.4), but occasional unusual displays have been recorded, as in 1803, 1922, and 1982. The 1982 outburst lasted about an hour, and was best observed from Florida. Here, experienced observers, including Norman McLeod, Paul Jones, and David and Brenda Branchett, saw rates of 75–80 Lyrids per hour on the night of April 21–22. The short-lived peak came at 0650 UT (solar longitude 32.1°), during daylight for European observers. By the

Figure 5.3 *Lyrid radiant position.*

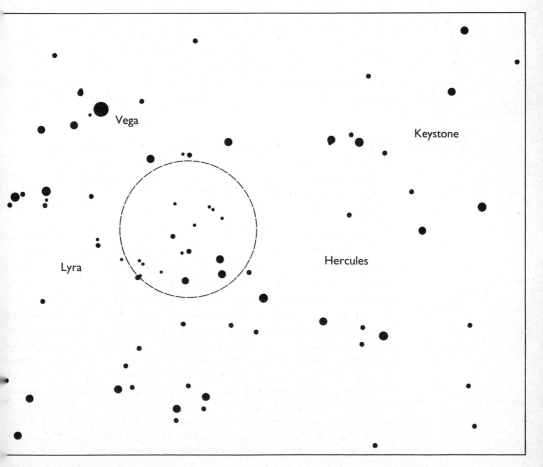

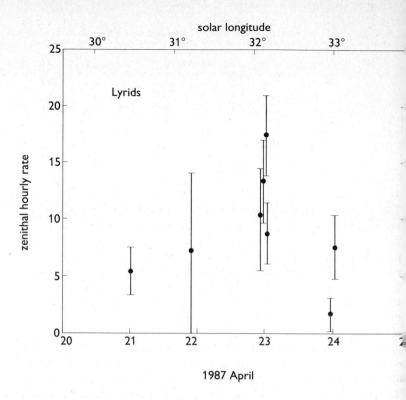

Figure 5.4 *Corrected ZHRs for Lyrid meteors observed in 1987. Peak occurred on the night of April 22–23, around 32.1° solar longitude.*

following evening, observed rates, while still respectable at 10 meteors per hour, had fallen back to normal peak levels.

For a time in 1982 the Lyrid ZHR touched 200, presumably as a result of the Earth running through a much denser region within the main meteor stream. The observation of this activity from only a limited part of the Earth's surface emphasizes the importance of having observers well spread in longitude: had no one seen the unusual activity in the early morning hours of 1982 April 22, we might now assume that the return had been typical of those in other years.

There are several other records of strong Lyrid returns. In 1922, an observer in Greece saw over 100 Lyrids in an hour. Similar rates may have occurred a number of times in the nineteenth century, the return of 1803 being particularly notable. Evidence has also been found for Lyrid activity at storm levels in oriental historical records.

Lyrid meteors originate from Comet P/Thatcher (1861 I), as was deduced by the German mathematician Johann Galle (1812–1910). The similarity between the orbital elements for the meteor stream and the comet made this link obvious:

	ω	Ω	i	e	a (AU)
Comet P/Thatcher	213.4°	30.3°	79.8°	0.984	?
Lyrids	214.3°	31.7°	79.0°	0.968	28

Table 5.3 *Lyrid radiant altitudes (in degrees), for various latitudes.*

| Local time (h) | Latitude. | | | | | | | |
	25°N	30°N	35°N	40°N	45°N	50°N	55°N	60°N
21	—	1.4	5.2	8.2	11.0	13.8	16.6	19.2
22	10.2	13.3	15.7	18.0	20.2	22.3	24.2	26.0
23	22.8	24.9	26.8	28.6	30.1	31.5	32.5	33.3
00	35.2	37.0	38.5	39.7	40.5	41.0	41.1	40.8
01	47.9	49.4	50.5	51.1	51.1	50.6	49.7	48.1
02	60.6	62.1	62.8	62.6	61.6	59.9	57.6	54.7
03	73.1	74.8	75.1	73.7	71.2	67.8	63.9	59.7
04	82.7	87.1	86.5	81.7	—	—	—	—

The comet has an orbital period of around 410 years, the longest of any known comet associated with a regular, annual meteor shower. The high inclination of the stream's orbit reduces the effects of gravitational perturbation by the planets. The stream is encountered at the descending node of its orbit by the Earth. Obviously, the high rates in years such as 1982 have nothing to do with the proximity of Comet P/Thatcher to the inner Solar System. Such strong returns happen when the Earth passes through denser "filaments" of meteoroidal material within the stream, released at perihelia of the parent comet in the far-distant past.

Although rates in most years are rather modest, the possibility of occasional and unpredictable stronger returns makes the Lyrids worthy of attention each year, provided moonlight does not hamper observations. The radiant attains its best altitude in the early morning hours (Table 5.3); Lyrid watches before midnight tend to be rather unproductive.

Eta Aquarids (April 24 – May 20)

Recent additions of material to that part of the zodiacal dust cloud through which the Earth travels have come predominantly from two short-period comets: P/Encke, the parent of the Taurids, and that most famous object, P/Halley. The stream of debris injected into the inner Solar System by P/Halley is encountered by the Earth twice in the course of the year.

The first encounter, 0.065 AU from the descending node of the stream's orbit, brings the Eta Aquarid shower, which peaks in early May. This is unfortunately a very difficult shower for observers at more northerly latitudes since the radiant (near the Water Jar asterism in Aquarius) is only just beginning to climb above the horizon haze as dawn breaks. Those at lower latitudes, including the southern United States, are better placed to observe these very swift meteors.

The orbits of both Comet P/Halley and the Eta Aquarid meteor stream are retrograde (in other words, the comet and meteoroids travel around the Sun in the opposite direction to the Earth and the other planets). Consequently, Eta Aquarid meteoroids impact almost head-on with the atmosphere at a geocentric velocity of 67 km/s (42 mile/s), and the high energies involved result in a high proportion (at least 25 percent) of the meteors showing persistent trains. The shower produces a reasonable crop of meteors in the fainter magnitude ranges. The brighter meteors are commonly yellow in colour.

For observers at lower latitudes, the Eta Aquarids are one of the year's most significant showers, producing observed rates of up to 30 meteors per hour, corresponding to ZHRs of between 50 and 60. A broad peak is found from about May 1–5, though the Earth actually comes closest to the "centre" of the stream's orbit (the *prime orbit*) on about May 8. This implies that meteoroids in the stream are dispersed some distance from the prime orbit. The presence of multiple peaks and slight troughs in Eta Aquarid (and, as we shall see later, Orionid) activity are taken to suggest a "shell" structure for the stream, built up from numerous filaments ejected from P/Halley's nucleus at different epochs.

The Eta Aquarids can be traced back in historical records to 74 BC. There are no indications that returns of the parent comet to the inner Solar System give rise to enhanced activity. Eta Aquarid observations made in 1985–6 under the auspices of the International Halley Watch revealed normal activity from the shower. The current activity of the Eta Aquarids is produced by debris ejected from P/Halley's nucleus in the far-distant past.

A second encounter with this stream, at its ascending node during the autumn, produces the Orionid shower which, as we shall see, is somewhat more accessible for observers at northerly latitudes.

Early summer

The arrival of summer and its welcome warmer daytime weather brings us to something of an off-season for meteor observers. There are only a few relatively minor showers active at this time, and observation of them is restricted somewhat by the shortness of the nights, particularly at higher latitudes. Indeed, around midsummer the sky never becomes properly dark at latitudes higher than about 50°. Some would suggest that this time is best spent taking a brief rest from observing, checking equipment and getting ready for the autumn onslaught.

There are, however, a couple of minor showers which merit the attention of observers at lower latitudes, where there are still a couple of hours of proper darkness in which productive watches can be conducted. As at other times of the year, there is a steady trickle of low activity from around the ecliptic. The *Alpha Scorpiids* are active from mid-April to mid-May, with a radiant just north of Antares. The ZHR is typically only around 5 meteors per hour, and with the radiant low in the southern sky this is not a particularly easy shower for northern observers. Long, slow meteors rather like the Virginids are produced, and these can sometimes be reasonably bright. Maxima may occur around April 28 and May 13.

From mid-May to about the end of June, activity from the southern part of the sky comes from the *Ophiuchids*, whose radiant lies near Theta Ophiuchi – about 15° east of Antares and 10° south of the second-magnitude star Eta Ophiuchi. Like the Alpha Scorpiids, the Ophiuchids produce only low rates and are not particularly favourable for more northerly observers. Peaks may occur around June 9 and June 19.

June activity in some years of the early twentieth century was augmented by periodic displays of the *Pons–Winneckids*, or June Draconids. This shower, which has not returned since 1927, was produced by a swarm of debris associated with the nucleus of a member of Jupiter's comet family. Comet P/Pons–Winnecke has an orbital period of about 6 years, and for a time gravitational perturbations moved the stream's orbit into a position where it could approach the Earth. This was forecast to happen in 1916, and led to a strong meteor display (observed rates of up to 100 meteors per hour) recorded by William Denning. The display was short-lived, as was a second return seen from Japan in 1921. The final observed return in 1927 was seen from Russia. Meteors from the Pons–Winneckid radiant were generally faint, and rather slow. Gravitational perturbations have now pulled the stream's orbit away from the Earth, and have also increased the perihelion distance of the parent comet, which is now a very dim object observable only in the largest telescopes, at best.

Around midsummer, the Earth's passage through the stream of debris left by Comet P/Encke gives rise to the *Beta Taurids*. This, unfortunately, is a daytime shower, and cannot be observed visually – though it may just be possible to detect isolated members of the shower in the growing dawn at the end of June. The shower was detected using radio methods at Jodrell Bank during the late 1940s, and has been found, together with the *Daytime Arietids* (active at around the same time), to be one of the strongest radio meteor showers of the year.

The *Alpha Cygnids* produce low rates, perhaps 1 or 2 meteors per hour, over a long period starting towards the end of June and continuing through to the end of August. As the name suggests, their radiant is close to Deneb, and is therefore high in the sky for northern observers. A curiosity about this shower is that its radiant seems to stay put, showing none of the drift normally observed with other showers. This has prompted the suggestion that a number of showers, coming "on stream" one after another, might share a common radiant by coincidence; this must be considered unlikely.

The southern showers

Peak season: July and August

As the nights begin to lengthen again after the summer solstice, so the pace of meteoric activity quickens. From mid-July onwards there is plenty to keep the observer busy. Good weather at this time of year can also be conducive to a concerted effort, and it may be possible to put in several consecutive nights' monitoring of the ever-changing activity from the southeastern part of the sky, where there is a complex set of radiants in Capricornus and Aquarius (Figure 5.5). Here is another example of activity from streams lying close to the plane of the ecliptic, subject to planetary perturbations, and split into multiple radiants. Several radiants are active simultaneously. This can make identification of individual meteors something of a challenge, and even experienced observers will have to check positions carefully. The combined activity of these showers can produce quite healthy observed rates of around 20–25 meteors per hour, particularly towards the end of July.

Capricornids (July 5 – August 20)

The first of the southern radiants to come to prominence in the summer is initially located at the western end of Capricornus, near the bright

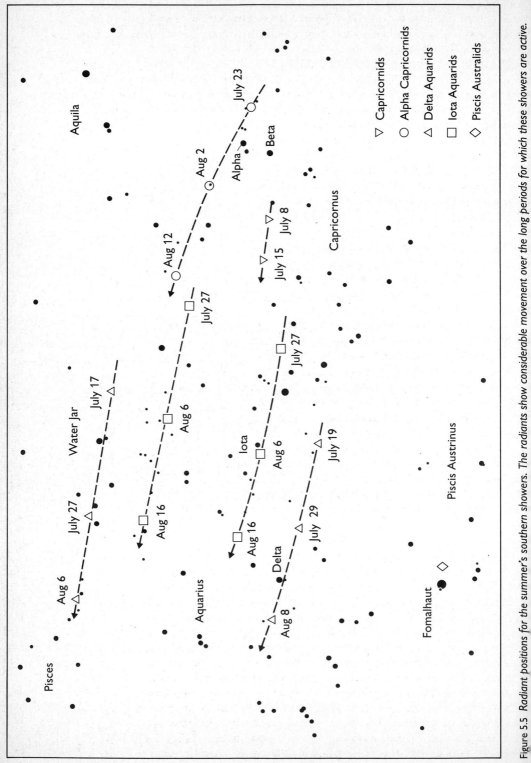

Figure 5.5 *Radiant positions for the summer's southern showers. The radiants show considerable movement over the long periods for which these showers are active.*

naked-eye pair Alpha and Beta Capricorni. As time goes on, the radiant drifts eastwards, and by the time of its second maximum on July 15 it lies 15° west of these stars, almost at the top centre of the wedge-shaped constellation. Capricornid meteors are slow, and sometimes bright. During watches in July I have seen a fair number of yellow–blue Capricornids, travelling upwards from the southeastern sky on long paths. On average, Capricornids are probably slightly brighter than the contemporaneous sporadic background.

Alpha Capricornids (July 15 – August 20)

Complicating the activity pattern in July is the presence, in the same constellation, of two radiants. In order to assess the relative activities of these showers, the observer has to be able to identify which meteors come from which sources – not always an easy task. The Alpha Capricornid radiant lies west of that of the Capricornids, some 5° to the northwest of Alpha and Beta Capricorni to begin with, and drifting eastwards by about a degree per day. At its peak, around August 2, the radiant is a few degrees northeast of Alpha Capricorni, and 10° west of the Capricornid radiant. Careful observation is required to identify meteors as belonging to either radiant (Figure 5.5).

Over the years the Alpha Capricornids have acquired a reputation as a shower producing reasonable numbers of bright meteors, and even a few fireballs. This does seem to be justified. Many regular observers have reported long, slow Alpha Capricornids climbing into the southern sky during watches in late July and early August, and observations in 1992 further confirmed that the shower can produce spectacular events. The brighter meteors are often yellowish, and show slow flares in brightness. One or two red Alpha Capricornids have even been seen.

Observed rates are usually never much more than 1 or 2 per hour and, like other showers whose radiants have large negative declinations, the Alpha Capricornids are best seen from lower latitudes. The low rates and preponderence of brighter meteors imply that the Alpha Capricornid shower is an old one, nearing the end of its evolution.

Delta Aquarids (July 15 – August 20)

Further complication is added to the activity from the southern sky by the presence of multiple radiants in Aquarius, the next constellation along from Capricornus (Figure 5.5). Most active among the showers emanating from this part of the sky during late July and early August are the Delta Aquarids. The Delta Aquarids have a double radiant, each one coming to its peak at a different time.

Of the two sub-streams, the southern is the more active. The southern radiant lies a couple of degrees south and west of Delta Aquarii at maximum on July 29 (solar longitude 126°). Observed rates from this radiant are often quite low for observers at higher latitudes because of its negative declination and consequent lower elevation. At best, an observer in the British Isles might expect to see 5–6 meteors per hour from the southern Delta Aquarid radiant, corresponding to a peak corrected ZHR of 20–25. US observers might expect to see rates about twice as high.

The northern Delta Aquarids arrive at maximum about a week later, on or around August 6, when their radiant lies near the Water Jar asterism. Observed rates from more northerly latitudes may be similar to those for the southern stream at its peak, but the resulting corrected ZHR is generally lower.

Meteors from both radiants are of medium speed, with a geocentric velocity of around 40 km/s (25 mile/s). Averaged Delta Aquarid magnitude distributions seem to be similar to those of the contemporaneous sporadic background: there are fewer bright events than from the nearby radiants in Capricornus. One or two brighter meteors are occasionally seen, mainly from the southern radiant. Brighter Delta Aquarids show yellow or yellow–blue colours. About 10 percent of Delta Aquarids leave persistent trains.

The shower is often well observed in those years when the Moon is badly placed for the Perseids – full moonlight in mid-August means that late July is favourably dark. Many observers look on the Delta Aquarids as something of a compensation in years when the Perseid maximum is "lost."

Iota Aquarids (July–August)

Like the Delta Aquarids, this is another ecliptic stream with a double radiant. The Iota Aquarids produce lower rates than their better-known neighbours, and are most notable for the relative abundance of fainter meteors. Iota Aquarids are generally swift, and emanate from radiants at the eastern end of Aquarius (Figure 5.5). The main peak is sometimes noted by early Perseid observers, around August 8. Rates are never particularly high: 2–3 meteors per hour is typical.

Piscis Australids (July–August)

Perhaps the least well-studied of the numerous radiants from the southern ecliptic region active during July and early August, the Piscis Australids produce low observed rates of only 1 or 2 meteors per hour. The radiant, a few degrees west of the first-magnitude star Fomalhaut (Figure 5.5), has a low declination, making this a very difficult shower for observers at higher latitudes, and few Piscis Australids are recorded from Britain or northern Europe.

Those few Piscis Australids which are seen appear to be slow with long paths, rather like Alpha Capricornids, and care should be taken to avoid confusing the two showers. Unlike the Alpha Capricornids, however, the Piscis Australids are not noted for bright meteors. The shower requires further study, and could usefully be observed from lower latitudes, such as the southern United States.

Perseids (July 25 – August 20)

There are few absolute certainties in meteor observing. One thing that does seem certain, however, is that nearly every amateur astronomer in the northern hemisphere seems to look on August as *the* time for meteor observing. Everyone likes to observe the Perseids – the great "free firework display" peaking around August 12. Even the general public is often aware that this is the time to see meteors. The high activity of the "August meteors" has been known since early times, and the Perseids can be traced back through the annals to the first century AD. In folklore, the Perseids are referred to as the "tears of St Lawrence," whose feast-day falls on August 10, close to the shower's maximum.

Their consistently high activity makes the Perseids appeal to even the most casual of meteor observers. The shower has the added advantage of coming during the summer holiday season, and often coincides with spells of fine weather and reasonably mild nighttime temperatures – in sharp contrast with the other two of the "big three" showers, which occur in the depths of winter.

The Perseids are a classic example of a cometary meteor stream. Their association with Comet P/Swift–Tuttle (the Great Comet of 1862) was established by Schiaparelli from analysis of orbital characteristics:

	ω	Ω	i	e	a (AU)
Comet P/Swift–Tuttle	152.8°	138.7°	113.6°	0.96	24.3
Perseids	151.5°	139.0°	113.8°	0.965	28

On the basis of observations obtained in 1862, the orbital period of P/Swift–Tuttle was taken to be 120 years, leading to the expectation that it would return – possibly attended by enhanced Perseid meteor activity – in 1981–2. Although good Perseid rates were seen at this time, the comet was conspicuous by its absence. Some astronomers suggested that P/Swift–Tuttle had swept past unseen. However, Brian Marsden, at the International Astronomical Union's Minor Planet Center at Cambridge, Massachusetts, proposed that the comet is identical with Comet Kegler of 1737, giving an orbital period close to 130 years – in which case it would return in the early 1990s. P/Swift–Tuttle was finally recovered by the Japanese observer Tsuruhiko Kiuchi in 1992, in late September, and its identity with Comet Kegler confirmed. The comet was a diffuse, difficult object in the early autumn skies, improving markedly before reaching perihelion that December. Many amateur observers in the northern hemisphere enjoyed fine views of the comet as a fifth-magnitude object in November.

The uncertainty in the precise orbital period of the comet arose from large *non-gravitational perturbations*, caused in particular by strong gas jets from the nucleus, which may significantly accelerate or decelerate a comet around the time of perihelion. The late return of Comet P/Brorsen–Metcalf in 1989 is a further example of how non-gravitational perturbations can affect a comet's orbital period. Since they vary from one comet to another, these effects are virtually impossible to forecast, and cometary orbital elements are consequently subject to some degree of uncertainty, with P/Swift–Tuttle being a particularly extreme case.

The Perseid meteor stream is encountered near the descending node of its retrograde orbit. The pattern of Perseid activity is normally reasonably consistent from one year to the next, though there is some variation in peak intensity. The first signs of activity appear around July 25, when the radiant lies in Cassiopeia, about an hour of right ascension from its maximum position. Observers should be aware of the radiant's daily motion during the relatively long period of activity, as shown in Figure 5.6.

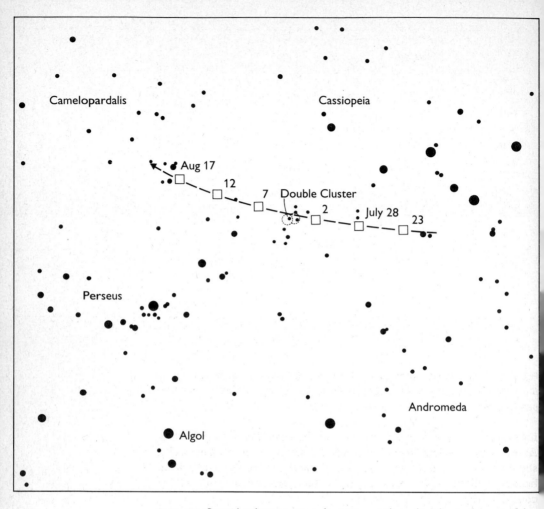

Figure 5.6 *Perseid radiant positions. At maximum, the radiant lies to the east of the Double Cluster.*

Early observed rates are low – only 1 or 2 meteors per hour. It is only as we enter early August that Perseid activity really begins to come to prominence. As shown in Figure 5.7, there is a steady, gradual rise in Perseid rates over the first week of August, continuing until about August 8–9. Then activity begins to rise more steeply, and it is in the next few nights that many amateur observers will reap their richest meteor harvest of the year. Between August 9–10 and 13–14, an experienced observer with at least five hours of clear dark sky available for watching may record upwards of 100 meteors on most nights. On maximum night, observed rates of 30–40 meteors per hour are not uncommon, especially towards dawn as the radiant climbs higher into the sky (Table 5.4).

The radiant at maximum lies close to the Double Cluster (h and Chi Persei) in northern Perseus – an easy naked-eye deep-sky object under good conditions. There may be other sub-radiants: it has been suggested,

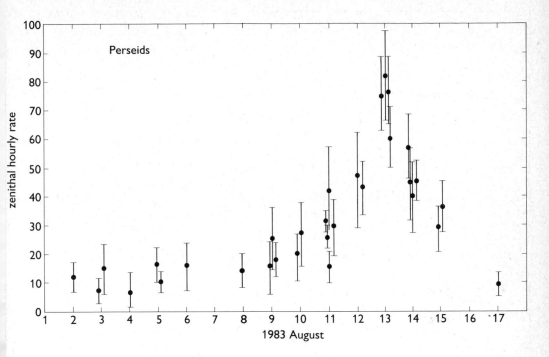

Figure 5.7 *Perseid ZHRs calculated from extensive observations in 1983, showing the peak around midnight on the night of August 12–13. The slow rise towards peak and the subsequent rapid decline are typical.*

for example, that early activity in August comes predominantly from an Alpha–Beta Perseid radiant. The high inclination of the stream's orbit should, however, limit the extent to which the Perseids can have become split into sub-streams, explaining why it has remained so compact over a long period of time.

In most years the peak of the Perseids (Figure 5.7) is fairly sharp, usually occurring on the night of August 12–13. The height of this peak has shown much variation in recent years. Throughout the late 1970s and the 1980s, peak Perseid rates were high but variable. It is worth

Table 5.4 *Perseid radiant altitudes (in degrees), for various latitudes.*

Local time (h)	25°N	30°N	35°N	40°N	45°N	50°N	55°N	60°N
20	—	2.9	7.6	12.4	17.1	21.8	26.6	31.2
21	3.1	7.6	12.1	16.7	21.2	25.7	30.1	34.5
22	9.2	13.6	17.9	22.2	26.4	30.6	34.7	38.7
23	16.3	20.4	24.6	28.6	32.6	36.4	40.2	43.7
00	23.9	28.0	32.0	35.8	39.5	43.1	46.4	49.5
01	31.9	35.9	39.8	43.5	47.1	50.4	53.3	55.9
02	39.7	43.8	47.7	51.5	54.9	58.1	60.7	62.8
03	46.9	51.2	55.4	59.3	62.9	66.0	68.5	70.0
04	52.8	57.4	61.9	66.3	70.3	73.8	76.4	77.5

noting that poor returns early in the twentieth century appear to be indicated by results obtained by Denning and his colleagues.

The Perseids were intensively studied by groups of observers in Scotland in August 1977, under the guidance of Robert McNaught. McNaught's team recorded unsurpassed numbers of photographic Perseid trails for analysis, in addition to acquiring a great deal of visual data. From these results, peak-corrected ZHRs of the order of 80 were found, which seems typical of activity during the late 1970s, and for much of the 1980s. The early 1980s, however, demonstrated that even a shower as supposedly well-understood as the Perseids can still spring the occasional surprise: exceptionally high maxima were observed in 1980 and 1981.

The maximum of 1980 was well observed by members of the BAA Meteor Section, participating in a specially organized project. Similar projects are frequently arranged to cover (weather permitting) most major showers on a national basis by the BAA and other organizations in years when moonlight is absent.

In what was generally a poor summer for observing in the British Isles, the 1980 BAA project was lucky to strike clear skies right on Perseid maximum itself. An Atlantic frontal system moved across the British Isles during the evening of August 12, leaving the skies brilliantly clear. Several observers can remember the frustration of apparently being rained off just before midnight turning to the joy of finding clear skies in the early morning hours. The moral of the tale is: don't give up on a key night – rain may mark the passage of a cold front, bringing better-than-normal skies.

For those who persevered until clear skies came along, the maximum of the 1980 Perseids exceeded all expectations. In Edinburgh, a group of three observers – Dave Gavine, John Sim, and Gerry Taylor – logged a remarkable total of over 400 meteors in a period of just four and a half hours. At Morpeth in Northumberland, Alastair McBeath and Kevin Blaylock each recorded similarly high totals. Perhaps most remarkable of all was the total of 903 meteors obtained by a team of 18 observers from the South Downs Astronomical Society in Sussex, who were clouded out until after midnight!

The maximum ZHR of the 1980 Perseids touched 120, higher than anyone could have predicted. Associated with these exceptional rates was a high proportion of fireballs and other bright meteors, giving astrophotographers a field day. This led to the suggestion that the Perseid activity was enhanced by the Earth encountering a filament of more concentrated, larger meteoroids within the Perseid stream. In most years, such "spines" are probably missed, and the peak is closer to the more "normal" ZHR of 80. In 1981, however, reports of similarly high Perseid rates for American observers appeared in the magazine *Sky & Telescope.* The very high rates in these years, which seem to have been confined to a narrow band of solar longitude, were recorded after dawn by UK and European observers in 1981.

The 1982 peak was badly affected by moonlight, but good rates were still seen. The 1983 return coincided with a very fine spell of weather for British observers, who amassed a very useful set of data. This time the return seemed normal, the peak ZHR on August 12–13 reaching

80. Coincidentally, observers at higher latitudes also had the distraction of a fine auroral display during their watches that night.

Observations in 1984 and 1985 were again affected by moonlight, but there were no indications of unusual activity. In the absence of moonlight, 1986 confirmed the suspicion that activity had returned to the levels seen through the late 1970s. Further observations through the late 1980s also bore this out – in particular, the 1989 return was favourable for British observers, producing good but not outstanding rates.

Perseid activity in the early 1990s has again been of much interest. Some observations from the late 1980s hinted at a double maximum in Perseid activity, with a new, second peak several hours *ahead* of the previously recognized August 12–13 maximum. Observations in the early 1990s suggested that this early peak is associated with meteoroids in the region of space around the returning Comet P/Swift–Tuttle.

The 1990 return of the shower was seriously affected by moonlight, though watches of limited duration suggested that maximum rates were more or less similar to those of previous years. Much more favourable moonlight conditions in 1991 encouraged many observers to make a concerted effort to observe the shower. By and large, rates were again pretty much in line with expectations. While cloud was a problem in some areas, many BAA observers were able to cover the maximum on August 12–13. Rates were comparable to those seen in 1983: a good, but fairly normal return of the shower. Among the meteors seen were several bright events, one of which was recorded photographically with the aid of a spectrograph.

Accounts from the United States of high activity in the early hours of August 12 appear to have been exaggerated. There can be no doubt, however, that the outburst of Perseid activity seen by observers in the Far East was remarkable. For a period of about an hour around 1700 UT, Japanese observers recorded Perseid rates of up to 450 meteors per hour. As with the 1980 outburst, this may have been produced by the Earth passing through a denser filament in the stream lying 11 hours ahead of the expected maximum.

The 1992 Perseid return was eagerly awaited, particularly as there were several forecasts of a recurrence of the high rates reported from the Far East the previous year. A major problem, however, was the presence of strong moonlight, limiting the accuracy of visual observations. Nevertheless, observers in Japan and eastern Europe again saw high rates in the interval between 1900 and 2000 UT – some three or four hours earlier than expected. While it is difficult to conclude much from visual observations made under such moonlit conditions (the Moon was only a day from full), forward-scatter radio observations (see Chapter 7) leave no doubt that unusual Perseid activity, again associated with the early peak emerging in the late 1980s, did indeed occur.

An excellent set of radio data covering the whole of the 1992 Perseid shower was obtained by John Mason in Sussex. His observations on August 11–12 revealed a series of very sharp, short-lived outbursts of high radio meteor activity associated with the Perseids, principally in the interval from 1840 to 2020 UT. A brief, strong burst at 1930 UT was detected by US radio hams, whose reports were collected by Joe Rao and Shelby Ellis. Ellis has monitored radio observations of the Perseids since the 1950s.

From such reports, and the subsequent recovery of P/Swift–Tuttle, it has been inferred that the early peak in Perseid activity in the late 1980s and early 1990s resulted from the Earth encountering meteoroidal debris close to the comet, and not yet incorporated into the main Perseid stream, which continued to peak as before around August 12–13. It remains to be seen how long this new peak will remain visible after the comet's perihelion passage, but observations in the years around this return of P/Swift–Tuttle will surely lead to a better understanding of the relationship between comets and meteor streams.

It is precisely because they can, occasionally and unexpectedly, produce outbursts of activity that meteor showers are so interesting to follow from year to year. Many general astronomy books give the misleading impression that we have learned all there is to know about such streams as the Perseids; the events of 1980, 1981, and 1991 certainly suggest otherwise.

After the maximum, Perseid activity falls off quite rapidly. In common with several other major showers, the decline is more rapid than the rise. Residual Perseid activity can be seen up to about August 20.

While the peak ZHR certainly seems to show considerable variation in the Perseids, other characteristics of this fine shower are more predictable. The meteors are fast: Perseid meteoroids have a geocentric velocity of 60 km/s (37 mile/s). Consequently a high proportion of all Perseids (about 30 percent), and of the brighter meteors in particular, leave persistent trains. As noted in Chapter 4 (Figure 4.7), the Perseids show some depletion in the fainter magnitude ranges relative to the contemporaneous sporadic background. Conversely, the shower is noted for its relatively high proportion of brighter events. As we shall see in Chapter 6, such events make the shower an ideal target for photographic work, which on the nights around maximum, in particular, is often very productive.

Early autumn

Following the excitement of the Perseids, September can seem something of a let-down. It is at this time, however, that sporadic rates reach their highest for the year, as borne out by BAA observations in the late 1980s. There is also a fair amount of minor shower activity, including another *Alpha Aurigid* radiant, with peaks around August 28 and September 15. Activity is fairly low, but spans the period from late August to early October. Like their counterparts early in the year, these meteors seem to be slow and bright. Photographic work at this time of year may actually be quite productive.

Piscids (September–October)

Throughout late September and into October, the good levels of sporadic activity are complemented by meteors from yet another near-ecliptic radiant complex. The Piscids produce low rates, typically 1 to 2 meteors per hour, with possible sub-peaks around September 8, September 21, and October 13. The precise activity pattern requires further study.

Piscid meteors are slow, and long-duration events are sometimes seen. Their magnitudes appear comparable to the contemporaneous sporadic background, and few very bright events are recorded. Activity in early September comes from near the fourth-magnitude star Delta Piscium (there are few bright stars at all in this area of the sky, south of the Square of Pegasus). Later in the month, Piscids appear mainly from a position some 10° southwest of Delta Piscium. In mid-October, the main source of Piscid activity lies a little to the southeast of another fourth-magnitude star, Eta Piscium, some 10° west of Aries.

Giacobinids (October 6–10)

Also known as the Draconids, from their radiant near Beta Draconis, the Giacobinids are seen only in certain years when conditions are favourable. Their parent comet, P/Giacobini–Zinner, is another of Jupiter's captures, having an orbital period of 6 years.

Theoretical studies first raised the possibility of meteor activity associated with the comet in 1915 and 1926, times of close approach to its orbit by the Earth. Denning recorded what appear to have been a handful of slow meteors from the radiant in 1915, but activity was far from exceptional. In 1926 the Earth arrived near the descending node of the orbit 69 days ahead of the comet; again, little out of the ordinary was seen, though J.P.M. Prentice did record a number of possible shower members.

Much more activity was seen by European observers in 1933, when the Earth arrived close to the comet's orbit 80 days behind the comet. In a period of 4½ hours centred on 2000 UT, 1933 October 9, rates of between 50 and 450 meteors per hour were recorded.

The 1939 encounter brought the Earth to the orbit's descending node well ahead of the comet, and no unusual activity was noted. The 1946 return, however, was favourable, and was well observed. Radio echo methods were successfully employed for the first time, from Jodrell Bank, to follow the activity. A peak of 170 echoes per minute was recorded on 1946 October 9 at 0340 UT. Visually the display was also impressive, despite strong moonlight. In Canada, Peter Millman was able to record the display photographically. The photographs taken by Millman showed the radiant to be compact, implying that the meteoroids had not yet had sufficient time to spread far around or across the stream's orbit.

Gravitational perturbations shifted the orbit of the comet and its meteoroid cloud away from the Earth for a couple of returns (1959 and 1965), then brought it back to a more favourable configuration in 1972. Many observers watched for unusual activity at the 1972 return, when the Earth passed through the descending node 59 days after the comet, but not much was seen at all, suggesting that the cloud of debris around the comet's nucleus remains fairly compact.

In 1978 the Earth passed the node well ahead of the nucleus, and in 1979 well behind it, and in neither year was any Giacobinid activity noted. The 1985 return was of greater potential interest, and observers were alerted to the possibility of activity well in advance. The Earth would pass the node 26 days after the comet, and there might be good activity.

Visual observations from Europe during 1985 were fairly unproductive. From Japan, however, significant numbers of Giacobinids were observed visually on 1985 October 8, around 0940–1000 UT. One Japanese observer recorded 39 Giacobinids in a 14-minute interval around 1000 UT, during daylight in Europe. Most of the meteors were apparently slow and faint.

In England, the Assistant Director of the BAA Meteor Section, John Mason, was operating his forward-scatter meteor radio system (see Chapter 7). As shown in Figure 7.3, Mason recorded reflection rates well in excess of the normal sporadic background flux in the 4-hour interval between 0700 and 1100 UT on 1985 October 8 – the same time at which the Japanese observers recorded their high visual rates. The peak Giacobinid flux was estimated to have occurred around 0935 UT, with rates in excess of 1000 meteors per hour saturating the receiver system. After 1010 UT, rates fell away rapidly.

Observations such as these allow the cloud of meteoroids around the nucleus of Comet P/Giacobini–Zinner to be mapped in outline, and help to place limits on its spread around the orbit. The characteristics of Giacobinid meteors tend to suggest recent release from the cometary nucleus. It is possible that the meteoroids impacting on the atmosphere from this swarm still retain some quantities of volatile material; this would account for the high frequency of long-duration persistent trains among meteors so slow (geocentric velocity only 20 km/s, 12 mile/s). In this respect the shower may bear comparison with the periodic southern April shower the *Pi Puppids*, produced by recently released debris from another of Jupiter's comet family, P/Grigg–Skjellerup. There are also some parallels with the periodic activity of the Leonids, and of the Pons–Winneckids of the early twentieth century.

In 1992 the Earth passed the descending node of the comet's orbit 177 days after the comet, and no activity was seen, the distance from the nucleus being too great. In 1998, the Earth will pass the node 44 days ahead of the comet, giving some possibility of Giacobinid activity. This will be an interesting time, indeed, for meteor observers, with the dual possibility of Giacobinid and Leonid storms within six weeks of each other!

Late autumn

Orionids (October 15 – November 2)

During early May, the Earth runs fairly centrally through the stream of debris left behind by Comet P/Halley in the inner Solar System. The same stream is again encountered, though at a greater distance from the prime orbit, at its ascending node during October, producing the Orionids (Figure 5.8). (For orbital elements, see p. 23.)

Although slightly less rich in meteor numbers than the Eta Aquarids, the Orionids have the advantage, particularly for northern hemisphere observers, of being better placed in the sky. The principal radiant, close to the Orion/Gemini border just southwest of the second-magnitude star Gamma Geminorum (Figure 5.9), rises around 23.00 local time, and is high on the meridian by dawn. Long meteor watches beginning after midnight can be rewarded by excellent Orionid totals, and are

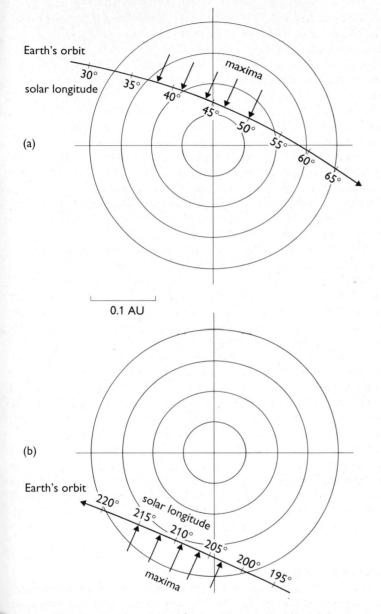

Figure 5.8 *Cross-sections through the P/Halley debris stream, showing the two annual passages of the Earth through it, producing (a) the Eta Aquarids and (b) the Orionids. The Eta Aquarid passage takes us closer to the stream's core, and gives slightly higher rates.*

much more productive than the brief glimpse of the Eta Aquarids in the gathering dawn in May. Table 5.5 gives the Orionid radiant's elevation as a function of local time around October 21.

Like Eta Aquarids, Orionid meteors are very swift, with a geocentric velocity of 66 km/s (41 mile/s). Persistent train phenomena are common,

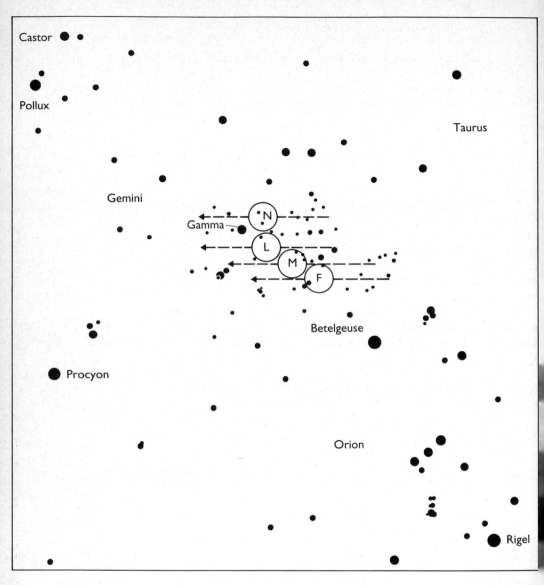

Figure 5.9 *Orionid radiant positions and motions. Four substreams – northern (N), leading (L), middle (M), and following (F) – were distingushed by J.P.M. Prentice.*

particularly among the brighter Orionids; recent studies have shown that some 40 percent of Orionids are accompanied by persistent trains. The figure is even higher if only the brighter meteors are considered – almost all Orionids brighter than magnitude 0 leave trains. The Orionids are on average slightly richer in bright meteors than the contemporaneous sporadic background. Figure 5.10 shows a comparison between Orionid and sporadic magnitude distributions.

The Orionids are active over a two-week period from mid-October to early November. The activity peaks in a broad plateau from October 20

Table 5.5 *Orionid radiant altitudes (in degrees), for various latitudes.*

Local time (h)	25°N	30°N	35°N	40°N	45°N	50°N	55°N	60°N
22	1.5	2.9	4.2	5.5	6.8	8.0	9.2	10.2
23	14.8	15.6	16.3	16.9	17.3	17.6	17.7	17.8
00	28.3	28.6	28.6	28.4	27.9	27.2	26.3	25.2
01	42.0	41.6	40.8	39.7	32.5	36.5	34.5	32.2
02	55.5	54.3	52.6	50.4	47.8	44.9	41.6	38.0
03	68.6	66.1	63.0	59.4	55.2	51.3	47.0	42.6
04	78.9	74.3	69.5	64.6	59.7	54.7	49.8	44.8
05	76.3	72.5	68.1	63.6	58.9	54.1	49.3	44.4

to 27, with several sub-maxima and minima superimposed upon it. Observers unlucky enough to hit one of the minima during their first Orionid watch may wonder what the fuss is about, but at best, around their primary peak on October 21, the Orionids can produce observed rates of up to 15 meteors per hour. Given the reasonably long hours of darkness available for observation, an Orionid watcher can clock up very respectable meteor numbers on a clear October night.

The broad Orionid maximum may be accounted for by activity from a number of sub-streams. During the 1930s, Prentice made a detailed study of the Orionid radiant and found a succession of radiants near Gamma Geminorum: northern, leading, middle, and following streams

Figure 5.10 *A comparison of magnitude distributions for Orionid and sporadic meteors during the 1980s.*

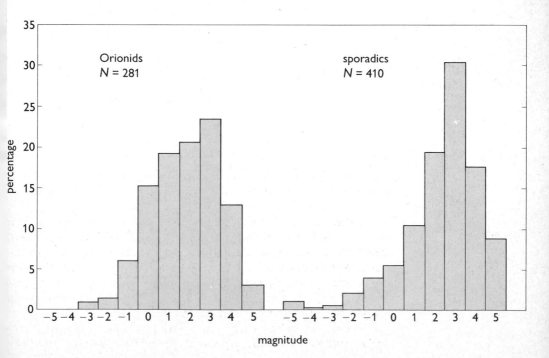

were all identified (Figure 5.9). More experienced observers may like to attempt an assessment of the relative activities of the various Orionid radiants, though extreme care is required for accurate identification.

Observations of the Orionids were part of the remit of the 1985–6 International Halley Watch, in which all phenomena associated with Comet P/Halley were studied in detail. There is no evidence to suggest that the return of the comet was marked by higher-than-normal rates from either the Orionids or the Eta Aquarids. The dust encountered at present was probably ejected from the comet at least a thousand years ago, and it will be several centuries before dust emitted at the 1985–6 perihelion passage disperses into the main stream. It is, however, important to keep this stream under observation whenever possible, since analysis of the various sub-maxima and how these may shift slightly from year to year can give us some idea of the evolution and distribution of dusty "filaments" and other structures within the stream.

The then BAA Meteor Section Director, George Spalding, carried out a detailed study of worldwide Orionid observations made in 1985 (Figure 5.11). He found peak ZHRs of around 24, in a broad maximum from solar longitude 208° to 210° (October 21–24). Activity was generally quite high (ZHR over 10) between October 19 and 26, with a possible secondary peak superimposed on the decline around solar longitude 213° (October 27).

Figure 5.11 *Orionid ZHR profile, based on results calculated by George Spalding from the 1985 return of the shower. A broad plateau maximum is seen around October 21–23, along with smaller maxima, and "troughs" in activity.*

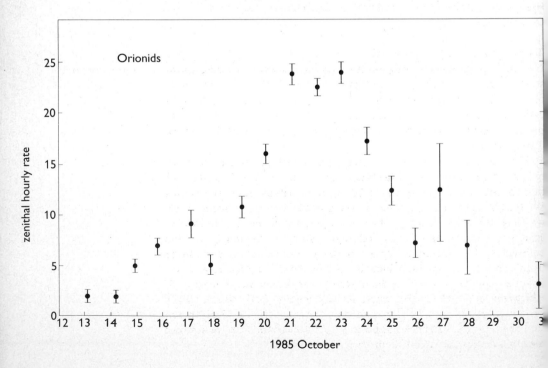

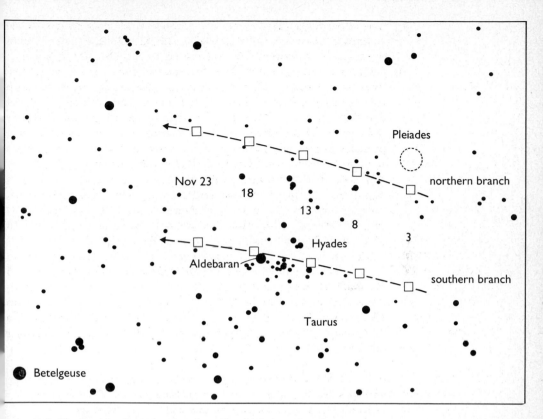

Figure 5.12 *The Taurid radiants (north and south), and their motions during October and November.*

Taurids (October 15 – November 25)

November is an important month for meteor observers, bringing, in some years, exceptional activity from the Leonids. In the late nineteenth century, the periodic Andromedids or Bielids were also a significant November shower (see Chapter 3). In most years, however, the principal meteoric attraction on late autumn nights is provided by the Taurid shower.

The Taurids are another shower whose meteors emanate from radiants lying fairly close to the plane of the ecliptic. The Taurids were first identified as a shower in the 1870s by observers, such as Thomas W. Backhouse, making gnomonic plots of meteors seen during their watches. It was not until the 1920s, however, that the double radiant structure became recognized following work by Denning and his colleagues, who included Alphonso King. Taurid meteors emanate from a northern radiant, which is initially near the Pleiades in late October, and a southern radiant which lies a few degrees west of Aldebaran and the Hyades when activity begins. Both radiants drift eastwards by about a degree each day (Figure 5.12), and observers should be careful to check the positions of the radiants before starting meteor watches during the Taurids.

Photographic studies of the Taurids during the 1930s allowed Fred Whipple to make the association between the shower and Comet P/Encke. The respective orbital elements are:

	ω	Ω	i	e	a (AU)
Comet P/Encke	185.2°	334.7°	12.4°	0.85	2.2
North Taurids	292.3°	230.0°	2.4°	0.86	2.59
South Taurids	113.2°	40.0°	5.2°	0.81	1.93

The difference in orbital inclination between meteor stream and comet can largely be accounted for by the gravitational influence of Jupiter.

P/Encke is a very short-period comet, returning to perihelion at intervals of only 3.3 years. The comet itself has an interesting observational history. It has been recorded on more returns even than P/Halley, and was independently discovered on several occasions by different observers. Charles Messier's colleague Pierre Méchain (1744–1805) was the first to find the comet, in 1786. He was followed by Caroline Herschel (1750–1848) – another member of England's most noted astronomical family – in 1795. The prolific French comet-hunter Jean Louis Pons actually discovered the comet twice at separate returns, in 1805 and 1818, but like P/Halley the object now bears the name of the mathematician who first surmised that the same body was being observed on successive returns, Johann Franz Encke (1791–1865).

Although it is now rather depleted in gas and dust as a result of its extremely short orbital period, Comet P/Encke was in the past responsible for emitting prodigious amounts of dust. Over time, this stream has

Figure 5.13 *A comparison of magnitude distributions for Taurid and contemporaneous sporadic meteors. The Taurids are proportionally richer in brighter meteors than the sporadic background.*

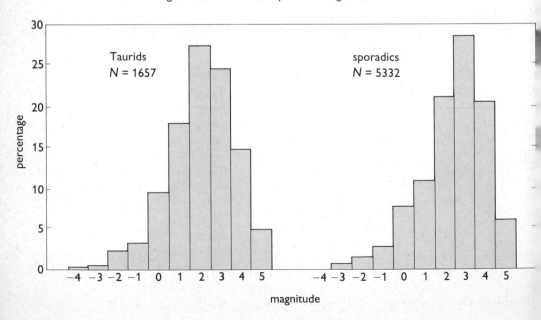

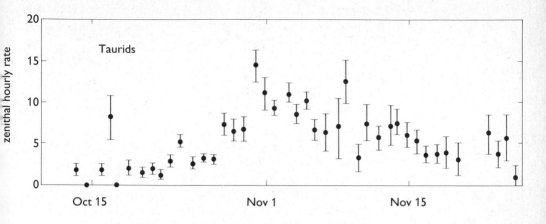

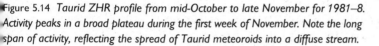

Figure 5.14 *Taurid ZHR profile from mid-October to late November for 1981–8. Activity peaks in a broad plateau during the first week of November. Note the long span of activity, reflecting the spread of Taurid meteoroids into a diffuse stream.*

spread out to fill a wide swathe of the inner Solar System. So extensive is the stream that all four of the inner (terrestrial) planets encounter it: Martian and Venusian observers could also see the Taurids.

The Earth encounters the stream twice during the year. The first encounter, in June, brings the daytime Beta Taurids. The later encounter during October and November gives a protracted spell of night-time activity, demonstrating the considerable spread of the meteoroids in this old stream. Whipple calculated that the meteoroids we see as Taurids in our atmosphere today were released as cometary dust about 4700 years ago. The presence of two sub-streams may result from periods of high dust emission at different epochs.

Historical studies indicate that activity from the Taurids during the eleventh century AD may have been comparable to today's Perseids. Gradual depletion by the Poynting–Robertson effect, and the spreading-out of the meteoroids by gravitational perturbations, have contributed to the shower's decline. There is evidence for the loss of smaller particles in the magnitude distribution (Figure 5.13).

The spread of Taurid particles is such that the Earth takes several weeks to traverse the stream. Observed Taurid rates are generally low – typically 1 or 2 meteors per hour in late October and from mid- to late November, and about 4–5 meteors per hour around the broad peak from October 31 to November 10 (Figure 5.14). These correspond to a maximum corrected ZHR of 10–12 meteors per hour. The low rates belie the fact that the Taurid stream actually contains perhaps ten times as much meteoroidal debris as the Perseid stream.

The proportion of Taurids leaving persistent trains is lower than for many other "cometary" meteor showers, typically 11 percent – about the same as for the contemporaneous sporadic background. This is presumably a result of their relatively low geocentric velocity, around 30 km/s (20 mile/s). Taurids are slow-moving, and sometimes of long duration.

Over the years, the Taurids have acquired a reputation for being a rich fireball source. Studies of BAA results obtained during the 1980s suggest otherwise, however, with only a small percentage of Taurids of magnitude -5 or brighter. It would appear that the low velocity and long duration of some *reasonably* bright (magnitude 0 to -2, perhaps) Taurids has sometimes led casual observers to overestimate their magnitudes. Photographic studies confirm that numbers of *very* bright Taurids are no greater than for the contemporaneous sporadic background.

The Taurids' reputation for being a rich source of very bright meteors may stem from rare occasions when a number of good events were seen, as in 1988 November, or on the night of a major, widely seen aurora on 1991 November 8–9. Overall, however, the regular observer will see only a small proportion of fireballs among the Taurids.

Leonids (November 15–20)

As we have seen in Chapter 3, this shower has played an important part in firing our interest in meteors and increasing our knowledge of them. The periodic storms have caused much excitement in the past, and it is likely that, with the possibility of a further strong return, all eyes will be on the shower in the late 1990s.

In common with many other showers, the Leonids have been misrepresented in the popular literature. It is wrong to state – as do all too many general texts – that Leonid activity is *absent* in those years away

Figure 5.15 *Leonid radiant position, in the Sickle of Leo.*

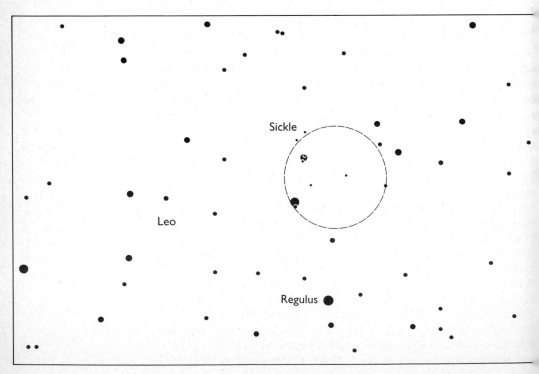

Table 5.6 *Leonid radiant altitudes ((in degrees), for various radiants.*

Local time (h)	25°N	30°N	35°N	40°N	45°N	50°N	55°N	60°N
23	—	—	—	0.0	2.5	4.9	7.3	9.7
00	5.1	7.0	8.8	10.5	12.2	13.8	15.3	16.6
01	18.0	19.3	20.5	21.6	22.5	23.2	23.7	24.0
02	31.2	32.1	32.7	33.0	33.1	32.8	32.3	31.5
03	44.7	45.1	45.0	44.5	43.6	42.3	40.7	38.7
04	58.3	58.1	57.2	55.7	53.6	51.1	48.1	44.9
05	71.9	70.0	68.6	65.5	62.0	58.1	53.9	49.6
06	—	—	—	—	—	61.8	56.8	51.9

(Column header group: Latitude)

from the perihelion of the parent comet, P/Tempel–Tuttle (1866 I). In fact, throughout the 1980s the few observers who made the effort were rewarded with quite respectable Leonid rates – up to 10 meteors per hour (better, indeed, than some average Taurid rates).

Popular texts also give a misleading impression of the great storms. Contrary to common belief, these stunning meteor displays are fairly short-lived and do not last the whole night. For example, the 1966 storm was at its peak for only about 40 minutes; activity on either side of this was, however, comparable to a good return of the Perseids, and, should similar behaviour be seen in the late 1990s, no one will be disappointed.

The identification of P/Tempel–Tuttle as the Leonid parent is credited to Schiaparelli. Orbital elements are as follows:

	ω	Ω	i	e	a (AU)
P/Tempel–Tuttle	172.6°	232.4°	162.7°	0.904	10.27
Leonids	172.5°	234.5°	162.6°	0.915	11.5

The Leonids radiate from the Sickle of Leo (Figure 5.15), the distinctive asterism shaped like a reversed question-mark with the first-magnitude star Regulus at its base. In order to see the Leonids, the observer must be prepared to wait up late: the radiant does not rise until about local midnight, reaching the southern meridian around dawn. Table 5.6 gives radiant altitude as a function of local time for various latitudes. As with other showers, observed rates are usually best when the radiant is highest. However, observers who delay starting their watches until the early morning hours in the key years of 1998 and 1999 run the risk of missing the short-lived peak.

Leonid meteors are very swift. Meteoroids in this stream have the highest geocentric velocity (70 km/s, 44 mile/s) known for any shower, close to the maximum value theoretically possible. The stream's orbit is retrograde, and is approached at its descending node by the Earth. A large proportion of these meteors leave persistent trains, some of which can be of extremely long duration. Leonids observed early in the night, just after the radiant has risen, sometimes follow very long paths across the sky.

The Leonids are rich in faint meteors, indicating a high proportion of small particles in the swarm. It is only because of their very high geocentric velocity that these small meteoroids produce enough energy to become luminous on impact with the Earth's atmosphere. Reasonable numbers of brighter events are also seen, as shown by, for example, Dennis Milon's classic photographs from 1966.

The cloud of meteoroidal debris (often referred to as the *ortho-Leonids*) around the nucleus of Comet P/Tempel–Tuttle will spread out over a period of centuries to fill the stream's orbit, as described in Chapter 1. Observations in the years leading up to the possible occurrence of great Leonid displays give some indication of the current extent of this spread. An important aspect of amateur studies of the Leonids during the early 1990s has been to check for the onset of slightly increased rates, which would herald the return to the inner Solar System of the outer fringes of the main swarm. Observations in 1991 November revealed nothing out of the ordinary; many students of the Leonids believe that we shall not start to see an upturn in rates until 1994. Many early-morning watches will be required in the closing decade of the twentieth century to check exactly how the swarm of debris surrounding P/Tempel–Tuttle is behaving. Observations in these years should also give some clues as to when, precisely, the possible storm can be expected. On the basis of the 1966 observations, maximum might be expected at a solar longitude of 235.16°.

Historically, the most significant displays have been in years when the Earth has passed the stream orbit's descending node after the comet, suggesting that the ortho-Leonid cloud lies principally behind P/Tempel–Tuttle in the orbit. The shower is still well worth watching in 1997, although we shall then pass the node 108 days ahead of the comet. Of greater potential interest, however, are the encounters of 1998 and 1999, respectively 257 and 622 days behind the comet (the 1966 storm occurred when the Earth passed the stream's ascending node 561 days behind the comet).

If the shower is most active at the same solar longitude as in 1966, peak might be expected around 1200 UT on 1998 November 17, or 1800 UT on 1999 November 17. These times would favour, respectively, observers in the western United States or the Far East. A strong return in 1998 would have to compete with moonlight: at the time of maximum the Moon will be 23 days old, and rising with the radiant. Conditions in 1999 will be more favourable, with an 8-day-old Moon setting by the time the radiant is above the eastern horizon.

Observations in the years leading up to the potential storm in 1998 or 1999 should help to refine the exact solar longitude of maximum activity, leading to better forecasts. While most observers avoid covering those showers affected by moonlight in a given year, an exception should be made for the Leonids. Results obtained even in the Moon-afflicted years of 1994 and 1997 will be of great value in determining the overall behaviour of this important shower.

It is difficult to estimate meteor numbers during a storm. One method used by observers in 1966 was to restrict counts to intervals of one minute, and count only those meteors appearing within certain areas of the sky, such as the Square of Pegasus. After the display, the counts were scaled up to yield values for the numbers of meteors over the whole sky. Naturally, during a storm it becomes impossible to record individual meteor details such as magnitudes and persistent trains.

The possibility of storm-level activity from the Leonids in 1998 or 1999, when the Earth will make its closest approaches to the swarm for 33 years, makes the shower a very important target for all observers. We can expect both amateur astronomers and the general public to show greatly increased interest in meteors around this time!

Geminids (December 7–15)   Winter

For many years an anomaly among the major meteor showers in having no obvious parent body, and meteoroids of higher-than-normal density, the Geminids are currently one of the three most active showers of the year – if not, in some years, the most active. The Geminids are often poorly observed as a result of poor weather, particularly in northwest Europe where early to mid-December can be marked by spells of foggy or cloudy conditions. The shower is, however, of considerable interest to both professional and amateur meteor students, and should be covered when at all possible.

Much of the professional interest stems from the rapid evolution of the stream's orbit. Computer models show that the Earth began to encounter the Geminids only in the nineteenth century, accounting for the absence of the shower from historical records. Over time, our annual passage through the Geminids has gradually brought us to different and richer parts of the stream. Even in the early twentieth century, the shower was apparently relatively poor, though the different observing techniques then favoured (as we saw in Chapter 3, most priority was given to meteor plotting) make it difficult to determine exactly what sort of rates the shower was then producing.

The theoretical studies by David Hughes at Sheffield University, and Iwan Williams and Kenneth Fox at Queen Mary and Westfield College in London, suggest that observations of the Geminids from early in the twentieth century should have revealed a different activity pattern, with the maximum early in the shower and a slow decline thereafter. Now, however, the Earth passes through the stream at a slightly different angle, as perturbations carry the Geminids outwards. In about another hundred years this effect will have been sufficient to drag the Geminid orbit away from the Earth's altogether.

Geminid meteoroids have always been regarded as unusual in having a much higher densy (2 g/cm^3) than those from "cometary" streams such as the Perseids (density 0.3 g/cm^3). An explanation for this apparent discrepancy was eventually provided by the discovery in 1983, with the IRAS satellite, of an asteroidal body – 3200 Phaethon – sharing the orbit of the Geminids:

	ω	Ω	i	e	a (AU)
Phaethon	321.7°	265.0°	22.0°	0.89	2.40
Geminids	324.8°	260.3°	23.6°	0.896	2.56

The Geminids' higher, "rocky" meteoroid density reflects the fact that they are fragments of an asteroid, perhaps crumbled away from its

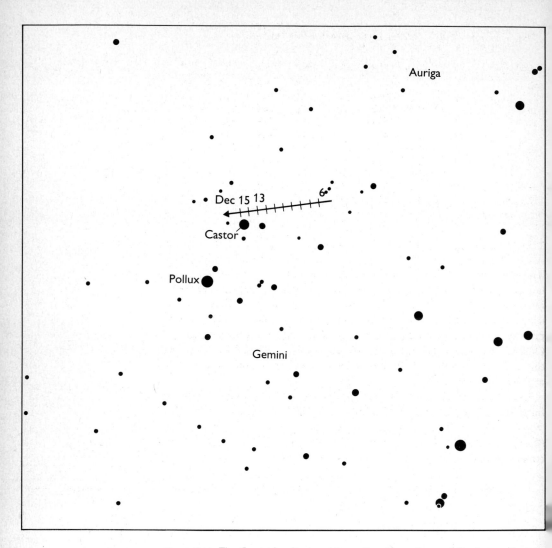

Figure 5.16 *The Geminid radiant and its motion. At maximum, it lies just northwest of Castor.*

surface by repeated heating at perihelion (Phaethon passes closer to the Sun, 0.14 AU, than any other known asteroid), or ejected following collisions with other asteroidal bodies.

As a result of the meteoroids' higher density and structural integrity, Geminid meteors often appear to last slightly longer before extinction than those from other showers. Brighter Geminids sometimes appear to break up into chains of luminous "blobs," all following the same trajectory. Their relatively low geocentric velocity of 35 km/s (22 mile/s) also adds to their slightly longer durations.

At maximum, the Geminid radiant lies slightly to the northwest of the first-magnitude star Castor, and is therefore easy to locate. As with other showers, there is a gradual eastward drift, as shown in Figure 5.16.

Table 5.7 *Geminid radiant altitudes (in degrees), for various latitudes.*

	Latitude							
Local time (h)	25°N	30°N	35°N	40°N	45°N	50°N	55°N	60°N
18	—	—	—	1.3	4.6	8.0	11.3	14.5
19	2.9	4.8	7.6	10.4	13.1	15.7	18.3	20.7
20	13.5	15.9	18.2	20.4	22.5	24.4	26.1	27.7
21	25.7	27.7	29.5	31.2	32.5	33.7	34.5	35.1
22	38.2	39.9	41.3	42.3	43.0	43.3	43.1	42.5
23	50.9	52.4	53.4	53.8	53.6	52.9	51.6	49.8
00	63.6	65.1	65.6	65.3	64.0	61.9	59.2	56.0
01	75.8	77.8	77.9	76.1	73.0	69.2	65.0	60.5
02	83.0	87.8	86.9	82.0	77.0	72.0	67.0	62.0
03	74.3	76.1	76.3	74.8	72.0	68.4	64.4	60.0
04	61.9	63.4	64.0	63.8	62.7	60.8	58.3	55.3
05	49.2	50.7	51.8	52.3	52.2	51.6	50.5	48.8
06	36.5	38.3	39.7	40.9	41.6	42.0	42.0	41.6

The Geminid radiant is favourably placed for observers in mid-northern latitudes, rising around sunset and staying above the horizon all night long (Table 5.7). The radiant culminates high in the south around 01.00 local time, and watches from late evening onwards can be very productive, particularly in the two or three days around maximum. Observers at lower latitudes, such as the southern United States, have the slight additional advantage that the Geminid radiant culminates at a greater elevation.

Like the Perseids, the Geminids currently build up slowly towards a fairly sharp maximum, followed by a rapid decline (Figure 5.17). The

Figure 5.17 *Geminid activity profile from 1991. Peak rates in recent years have been comparable to those of the Perseids.*

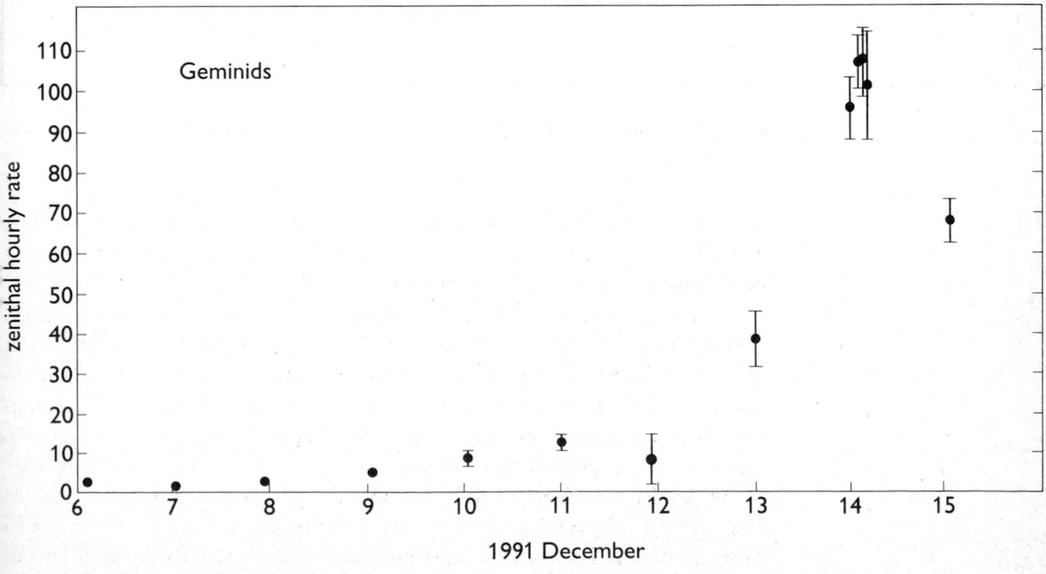

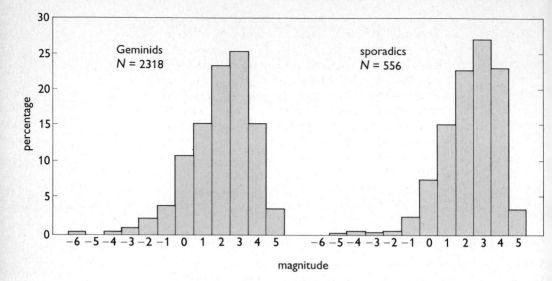

Figure 5.18 *A comparison of magnitude distributions for Geminid and sporadic meteors.*

first hints of activity are seen around December 6, but it is only around December 9–10 that rates begin to match, then overtake, the contemporaneous sporadic background. Maximum is usually on December 13–14, and can produce rates in excess of those from the Perseids in a normal year. In 1990 and 1991, experienced observers were recording Geminids in bursts of up to 3 or 4 per minute at maximum – almost too many to record accurately! Peak observed rates of 40–60 meteors per hour were recorded in those years, corresponding to corrected ZHRs of around 100.

From the computer-modelling studies carried out at Queen Mary and Westfield College and Sheffield University, it appears that the occurrence of peak Geminid rates at a later time should become apparent over the coming decades, and also that the peak rates themselves will increase with time. The Earth should pass through denser parts of the stream as these are pulled gradually outwards, towards the ascending node of the orbit, by gravitational perturbations. While observations from the early part of the twentieth century provide few clues, it does appear that observers enjoyed higher peak Geminid rates during the 1980s than during the preceding decades. Typical returns of the 1970s gave Geminid ZHRs at around 70 at most, but the returns of 1980 and 1982, both of which were well covered by BAA observers, produced peak ZHRs of 75–80, and in 1985 and 1988 rates may have peaked slightly higher. The high peaks of 1990 and 1991 were, according to some, a continuation of a steadily increasing trend in the shower. It will be of interest to see whether this presumed increase continues into the twenty-first century.

Magnitude distributions for the Geminids show the shower to be quite rich in bright events (Figure 5.18). The proportion of brighter meteors at a given time appears to depend on solar longitude, as shown by an authoritative analysis of 14,000 Geminid meteor magnitude estimates from the 1970s and 1980s carried out by Spalding. Bright Geminids become more common, in proportion, after the shower's visual maximum; the high rates which gave a ZHR approaching 100 in the early 1990s were produced mainly by meteors between magnitudes +2 and +4.

Relatively few Geminid meteors produce persistent trains, presumably because of their greater integrity and low geocentric velocity. Typically, around 3 percent of Geminids produce trains, compared with 5 percent for the contemporaneous sporadic background.

Their low velocities make the Geminids an excellent target for photographic work, and those who made the effort to obtain photographs in 1990 and 1991 were particularly well rewarded. The results can sometimes be spectacular – Geminid meteors occasionally show long, slow flares along their paths.

Ursids (December 19–24)

Coming just before Christmas, the Ursids are often neglected by observers in Europe and North America. As with the Geminids and Quadrantids, the frequently poor observing conditions at this time are also a hindrance. While the generally quoted peak ZHRs of 5–10 may also make this seem a shower of little observational interest, occasional outbursts of better activity can occur.

The Ursids first came to prominence in 1945, when observers in eastern Europe apparently detected high rates on December 22. Observers in the British Isles were surprised to record rates of up to 15 meteors per hour, corresponding to a corrected ZHR of as much as 45–50, on 1982 December 22–23. Four years later, in 1986, observers in Europe again noted reasonable Ursid rates.

The radiant, just north of Beta Ursae Minoris (one of the Guardians), is circumpolar for observers at northern latitudes, and is at a favourable altitude for most of the night. Ursid meteors are fairly swift, and have a magnitude distribution comparable to that of the contemporaneous sporadic background. A fair number of Ursids leave persistent trains, and the brighter meteors may show a yellow colour.

The shower is produced by debris from Comet P/Tuttle, encountered near the descending node:

	ω	Ω	i	e	a (AU)
P/Tuttle	207.0°	269.8°	54.6°	0.821	5.7
Ursids	205.9°	270.7°	53.6°	0.85	5.7

Comet P/Tuttle has an orbital period of 13 years, and may attain sixth magnitude at best, brighter than most other comets (see Chapter 8). The strong returns of the Ursids in 1982 and 1986 cannot be ascribed to debris close to the parent comet. Perhaps it is more likely that, like the Lyrids, the stream has some denser filaments running through it. Although rates in most years are close to the rather modest "textbook" levels, the Ursids are worth watching whenever possible in case they are more active than usual.

New showers?

From time to time reports appear of outbursts of activity from previously unrecognized radiants, raising the possibility that new meteor showers may be becoming active. While it is possible for the Earth to encounter new streams over the long term, one has to treat "new shower" reports with some caution. Many reports of new shower activity are made on the basis of limited observations – two meteors with an apparently common radiant, observed on successive nights, do not necessarily make a new shower. Reports of new showers need to be followed up over several years; most turn out to be spurious. If every apparently shared radiant were to be accepted, we would have showers emanating from every square degree of the sky on every night of the year, albeit most of them with ZHRs of zero for most of the time!

It can happen that outbursts of several meteors, apparently from a common source, can be seen on a given night. Many regular observers will have seen these, and wondered whether a new shower is emerging. Under such circumstances it may be worth plotting the meteors on a suitable chart for later analysis, ensuring that all relevant details are noted down. Such reports require verification, and observations will need to be made in subsequent years. Sadly, many such outbursts, which may very well be genuine, turn out to be one-off events – perhaps due to chance encounters between the Earth and very minor, under-concentrated streams which produce activity only in certain years.

As we have seen with the Giacobinids and Pons–Winneckids, such *periodic showers* do exist. These cannot always be forecast. Conversely, many potential showers associated with close approaches to the orbits of periodic comets fail to materialize – though negative observations are still of considerable value in these instances, as they show that there is no debris distributed in certain parts of the orbit. For example, in 1987 some astronomers suggested that a close passage of the Earth to the orbit of Comet P/Nishikawa–Takamizawa–Tago would give rise to enhanced activity from the Epsilon Geminids. This is a very minor stream, barely distinguishable from the sporadic background in most years. Alas, in 1987 it remained so.

Such disappointments aside, it is still worth keeping an eye on the current literature for forecasts of possible activity associated with such one-off cometary encounters. Earth-approaching asteroids could be another source of meteor activity, following the precedent set by 3200 Phaethon and its apparent relation to the Geminids.

One longer-term forecast, issued in 1986, points to the possibility of activity around 1997 November 9 from debris associated with Comet P/Hartley (2). The comet itself will pass fairly close to the Earth, and is expected to become a naked-eye object in October of that year. Meteors from the shower would have a radiant near Epsilon Pegasi, and would be slow, with a geocentric velocity of around 16 km/s (10 mile/s).

The continually changing pattern of meteor activity provides sufficient work for a lifetime of observation by the amateur astronomer. The simple naked-eye visual observing techniques described in Chapter 4 are adequate for most purposes. More advanced observing methods can also be applied, and these I discuss in the chapters that follow.

CHAPTER 6

Meteor photography

Many amateur astronomers complement their visual observation of meteors by photography. Meteors are relatively easy to photograph, and you can use just as much equipment as suits your purse and purpose. The results can often be spectacular, and are useful in a number of ways. Visual plotting has by and large been discarded by most amateur meteor observers as too inaccurate to allow precise positional work – not so with photography!

The rapid light fluctuations of a bright meteor may only be glimpsed by the visual observer, but will be recorded in faithful detail by a good camera: light-curves obtained from such work can sometimes reveal much about a meteoroid's structure and conditions in the upper atmosphere. Figure 6.1 shows a Perseid with a small flare right at the end of its path, caught at 2240 UT on 1983 August 12–13.

In its simplest form, meteor photography consists of setting up a camera on a sturdy mount aimed at the sky, and carrying out time

Figure 6.1 *Photography often reveals subtle light variations in meteors. This Perseid, photographed by the author at 2240 UT on 1983 August 12–13, shows a tiny flare right at the end of its path.*

exposures with a fast film: effectively the film emulsion acts as the camera's retina, and the camera is left to perform a "watch," the permanent record of which is subsequently processed and analysed at leisure. Naturally, as in all fields of astronomical observing, there is slightly more to it than that. For instance, not all cameras are suitable for this work, and neither will most nights of the year be as productive as those around shower maxima. Also, there is the element of good fortune. I was lucky with my own earliest attempts at recording meteors on film: on my first film, I recorded two, from successive nights in 1981 July. But then I had to wait 13 months for my next meteor trail. Meteor photography will certainly teach you patience, if visual observing has not already done so.

Equipment

Many modern cameras meet the two main requirements for meteor photography. First, the camera must have a B or T shutter setting, allowing long time exposures to be taken. This requirement rules out compact 35 mm cameras, which have a single, short exposure time, or only a small range. Single lens reflex (SLR) cameras usually come with the necessary shutter settings. A problem with the increasing automation of such cameras, however, is that long time exposures drain the batteries, so a camera with a manual operation option is preferable. Medium-format cameras, which take 120 film, can also be used. While the top-range models in this format are for most of us prohibitively expensive, many meteor observers have enjoyed considerable success with cheap medium-format cameras such as the German Lubitels or the US Seagulls. Each format has its devotees, and there are good cases to be made for both.

The second requirement is that the camera be equipped with a reasonably fast lens, which works well at quite low f-numbers. (The f-number is simply the ratio between the focal length of a lens and its aperture.) For meteor work, it is best to use camera lenses with fixed focal lengths. Zoom lenses should be avoided in astrophotography since, with the lens pointing skywards, there is always the risk that during the exposure the zoom mechanism could slip and defocus the image.

The camera lens aperture is altered by opening or closing an iris diaphragm within the lens assembly. Closing the diaphragm reduces the aperture of the lens, increasing the f-number, and reducing the amount of light transmitted to the film. The effect is similar to that of introducing aperture stops behind a refractor's objective lens to reduce its effective aperture and light-gathering power. For meteor photography it is generally desirable to work below $f/5.6$, as higher f-numbers give a rather narrow aperture and only very bright meteors will be recorded. It is not always possible, however, to work with the lens fully open, as will be explained later.

Other essentials are a tripod, or some other support to hold the camera steady during the exposure, and a locking cable-release. Portable tripods with telescopic legs are ideal, but you must use one that is sufficiently sturdy to stay steady in a breeze. There are two

Figure 6.2 *Simple camera set-up for meteor photography. The camera, with a wide-angle lens, is mounted on a tripod. A lens hood acts as a partial dew-cap, and a cable-release is used to keep the shutter open during long time exposures.*

principal types of portable tripod on the market. In one type, tightening screw-rings holds the legs in place, while in the other, locking tabs fulfil this function. In my experience, locking tabs are easier to work with in the cold. The locking cable-release is essential for keeping the shutter open for long periods, and can be easily – and relatively cheaply – obtained from a camera shop. Metal- or plastic-wrapped types are less susceptible to the damp than the cloth-wrapped type. Figure 6.2 shows a typical single-camera set-up which I use for meteor photography.

Film

Astrophotography takes advantage of the ability of films to build up progressively stronger images of faint objects as the exposure goes on – up to the limits imposed by low-intensity reciprocity failure (basically, the inability of the emulsion to convert excited silver ions to a stable form at low light levels). Thus, if a camera is guided to follow a starfield across the sky for a long period, images of faint stars, invisible to the naked eye, will be built up. The rate at which the images build up is determined by the *speed* of the film, indicated by its ISO rating. A film rated at ISO 400, for example, will record faint stars more quickly than one rated at ISO 100, and is therefore said to be faster.

In meteor photography, it is important to use a reasonably fast film since the light from even a bright meteor, as it streaks across the sky, will be focused on the emulsion for only a very short time. For this reason, most meteor observers use ISO 400 films, particularly Ilford HP5, or Kodak Tri-X or TMAX400. These are all black-and-white films, which have several advantages over colour film for meteor work: they are much less expensive, and processing is also cheaper and simpler, and can be done at home. Black-and-white films are better suited than colour films to push-processing, a technique for increasing effective speed and contrast, and which, under certain circumstances, increases the chance of obtaining clear meteor trails.

Colour film *can* be used for meteor work, but the results seldom justify the expense. Few meteors show a marked colour on film (they are not focused on a given area of the film for long enough to excite the colour layers in the emulsion), and trails may be lost in the heavier background grain.

Practice and problems

The simplest method of meteor photography is to make long time exposures – typically 10 to 15 minutes – and hope that a meteor sufficiently bright to record will appear in the camera's field of view during the exposure. The camera need not be driven: indeed, an undriven camera makes life easier for the observer. On the eventual image the stars will leave curved trails as they are carried across the field of view by the Earth's rotation, while any meteors recorded will appear as longer, straight trails cutting across the curved ones, and quite often showing changes in brightness. Figure 6.3 is a typical example, which I recorded in a 5-minute exposure with the set-up shown in Figure 6.3 on 1984 April 24–25.

Even under optimum conditions with a lens operating at $f/2$ and a fast film, only meteors brighter than about magnitude +1 will be recorded photographically. Faint meteors, by far outnumbering the bright ones, elude the camera and remain the property of the visual watcher. While this inability of photography to record all but the brighter meteors might appear to be a drawback, we can actually turn it to our advantage and use it to assess the activity of the brighter sub-population within a shower. For example, hourly rates for bright meteors in December's Geminid shower peak somewhat later than those for fainter meteors. The effect is clearly revealed when one compares photographic and visual rates: the visual peak, made up mostly of meteors of magnitudes +2 to +4, comes 6 hours or so before the peak photographic rates.

Figure 6.3 *A flaring sporadic meteor, photographed by the author at 2246 UT on 1984 April 24–25 during a 5-minute exposure using the set-up shown in Figure 6.2.*

Figure 6.4 *Satellites can often leave convincingly meteor-like trails. Shown here is a long, bright trail produced by the Mir space station.*

Like the visual observer, the would-be meteor photographer should keep a careful note of timings, recording the start and end times (UT) of each exposure for later reference. During the exposure, it is a good idea to keep a visual watch on the camera's field of view, recording details of all meteors seen. Note should also be made of any passing satellites or aircraft in the camera's field, since these can often leave convincingly meteor-like trails on the film. Figure 6.4 shows the trail of the Mir space station in July 1989, while Figure 6.5 shows the pattern produced by the lights of an aircraft crossing the northern sky.

Satellites in particular can wreak havoc with meteor photography during the favourable Perseid shower in August. At this time of year, the Sun remains above the horizon at orbital altitudes at the latitudes of northwest Europe and the northern United States, and at times the illuminated satellites traversing the observer's field of view seem to outnumber the meteors. Tumbling or rotating satellites may suddenly brighten or dim as a reflective surface or solar panel is turned towards or away from the Sun. Satellites brighten on emerging from the Earth's shadow, or fade from view as they are eclipsed by it. All these effects can combine to produce meteor-like trails on the film, so keep a good lookout. Many years ago the astronomer Walter Baade coined the term "vermin of the skies" to describe asteroids, having found many of their short trails on a deep-sky plate taken with a large telescope for astrophysical research. Perhaps the time has come to transfer the tag to the mass of junk in orbit which is increasingly contaminating the night skies.

One of the major problems in meteor photography is sky haze, particularly near large towns. Although the sky might often appear perfectly

127

Figure 6.5 *Aircraft, like satellites, are a source of unwanted trails on films exposed for meteor work. Alternating flashing aircraft lights often leave a trail of "footprints" across the negative.*

clear and dark, there are usually large amounts of suspended particles and water droplets present which tend to reflect and scatter city lights. Together with the all-pervasive light pollution from sodium lights, this suspended material is the cause of the orange hue often seen in, for example, deep-sky photographs taken from or near cities. Similarly, in meteor photography the sky haze tends to increase the background light level on the negative so that trails of fainter stars and meteors often become swamped.

There are two possible solutions to the sky haze problem. Either exposure times can be reduced, so that the camera is still used at high speed, or the *f*-number can be increased and the exposure time left as normal. The latter is less expensive on film, but will reduce the likelihood of capturing meteors. In the end, the observer has to decide to what degree successful meteor photography is worth the cost in film.

Success in meteor photography is influenced by a number of factors, few of which are under the observer's control. Perhaps the most obvious is the general level of meteor activity. The capture rate will normally be highest during the major annual showers, notably the Perseids and Geminids. This does not mean that it is not worth trying at other times, and it should be borne in mind that, if the change in the flux of bright meteors from night to night is to be monitored accurately, photographic work by dedicated observers needs to be carried out at all times. If, however, the observer's principal aim is simply to record meteor trails on film for their own sake, then the spring months, when rates are at their lowest, are probably best avoided.

The Perseids have been my main source of photographic success. Figure 6.6 shows a magnificent example from this shower, a magnitude −5 meteor which appeared at 0051 UT on 1983 August 13–14. This multiply flaring fireball lit up the ground and left a persistent train which lasted for 10 seconds. Later that same night, at 0134 UT, the Perseid shown in Figure 6.7 was recorded. This magnitude 0 meteor showed a

Figure 6.6 *A spectacular, long, flaring, magnitude −5 Perseid photographed at 0051 UT on 1983 August 13–14, using HP5 film at ISO 400 and a 29 mm f/2.8 wide-angle lens.*

Figure 6.7 *Another meteor from 1983 August 13–14, photographed at 0134 UT. A satellite trail cuts across the top right corner of the frame.*

Figure 6.8 *A magnitude −2 Geminid passes through the top end of Orion on 1990 December 12–13.*

smooth terminal flare; a satellite trail is seen to the right of the meteor. December's Geminids are also favourable for photography. I was very fortunate to record the flaring magnitude −2 Geminid shown in Figure 6.8 at 0009 UT on 1990 December 12–13, during a short break in the clouds.

It has long been known (see Chapter 4) that, for a given shower, some parts of the sky are richer sources of meteors than others. Fewer meteors are seen if the observer's attention is concentrated around the radiant, or at the zenith, or near the horizon: the optimum viewing direction is 50° altitude, and some 40° in azimuth from the radiant. Such factors also apply to meteor photography. It is usually preferable to aim the camera at 50° altitude, and 20–30° in azimuth from an active shower radiant. This displacement in azimuth is shorter than for visual observation because the shorter apparent paths of shower meteors close to the radiant give the appearance of slower meteor motion, and it allows longer for each point along the meteor's trajectory to excite the grains of the photographic emulsion than would be possible for the longer, apparently faster meteors farther from the radiant.

Even when a meteor shower is expected, successful meteor photography still depends on weather conditions. Clouds present obvious problems, especially when illuminated from below by streetlighting. While a visual observer can divert his or her gaze to a different part of the sky to avoid passing patchy cloud, this is impracticable for a fixed camera. Illuminated cloud passing through the field of view will soon fog the frame, and exposures usually have to be terminated immediately cloud shows up.

A major problem, and one often encountered on those clear, calm nights in July and August which are otherwise so favourable for meteor photography, is dew condensing on the lens. Dew can be avoided to some extent by keeping the camera well above grass-level. A lens hood, extended – just like a refractor's dew-cap (see Figure 6.2) – is helpful. As a last resort, a hair drier can be used to blast moisture off an already bedewed lens, but I need hardly emphasize the risk in using electrical equipment in damp outdoor conditions. A safer remedy is simply to take the dewed-up camera indoors to warm up again for a while.

Some observers use lens heaters, made from small electrical resistors giving out about 3 watts of heat when connected to d.c. torch/flashlight batteries. While this can be expensive on batteries, it does save on lost exposure time. The resistors, arranged in a ring around the lens (and obviously out of its field of view), generate heat which causes air to rise above the lens, preventing dew from forming on it.

If there is a single way in which the observer can influence capture rate to some degree, it probably lies with choice of camera lens. Modern SLR cameras have the great advantage of taking a range of interchangeable lenses of various focal lengths. In particular, wide-angle lenses can be useful in meteor work, provided all the observer wishes to do is obtain plenty of trails for their own aesthetic value, and for assessing rates. The scientific value of such an approach is limited, but we shall look later at the sort of equipment required to obtain meteor photographs suitable for precise measurements.

Most SLR cameras come with a 50 mm focal length lens, capable of operating at $f/2$ or faster as standard. This provides a field of view which will just neatly encompass, say, Cygnus. Wide-angle lenses of 28 mm focal length are readily available, and provide a much greater field of view – capable of taking in the whole Summer Triangle, for example. My experience with meteor photography has been that a wider field of view, even with a slightly slower lens, is a great advantage if all one wishes to do is capture meteor trails on film. Directly comparable results from exposures made during the Perseids in 1982 and 1983 show an average capture rate of one meteor trail every 1h 45m for a 50 mm focal length lens, and one every 1h 10m for a wide-angle lens.

Another advantage, from the aesthetic viewpoint, of the wide-angle lens is that the field is more likely to contain meteor trails in their entirety: smaller fields of view often result in only the beginning or end of trails being recorded.

Very short (9–16 mm) focal length fish-eye lenses have been used by some observers for meteor photography. At first glance these might seem ideal – what wider field of view could one wish for than the whole sky? Unfortunately, these systems (unless one is prepared to pay a great deal of money) operate at high f-ratios, and will capture only the very brightest meteors of all, the comparatively rare fireballs. As we shall see, however, such camera systems do have their uses in more serious work.

The best solution, then, for the fairly casual meteor photographer wishing to obtain "snapshot" trails on film is to use a reasonably fast wide-angle lens, ISO 400 black-and-white film, and concentrate on the maxima of the major showers.

In the darkroom

So, successful meteor photography depends on a number of factors including activity, choice of lens, and a spot of luck. The observer will certainly have invested a fair amount of effort by the time he or she arrives at the stage of having finished a film. It is important that care is taken at this point not to waste the work which has gone before, as film processing is as important to success as the factors mentioned above.

Most photographic shops have increasingly come to concentrate on colour processing and printing, but it is still possible to have black-and-white negatives developed commercially. Many meteor photographers prefer to process their own films, though, and there are sound reasons for doing so. For a start, not all (or, being realistic, sometimes not any!) frames on a film will contain meteor trails. There is little point in having meteor-free frames printed, as will be done as a matter of course by a professional processor, who will charge for all the prints. By developing the film yourself, you can at least have some control over which frames to have printed, even if you are unable to do your own printing.

Black-and-white film processing is relatively straightforward, and the necessary chemicals can be purchased reasonably cheaply from photographic suppliers. Using a commercially available developer allows the observer to control contrast and film speed; by adjusting development time, the film can be "pushed" to higher speed and contrast, improving the chance of more obvious meteor images coming out on the negatives. Developers in common use include Ilford ID11 and Microphen, and Kodak D19 and TMAX.

Processing takes about 20–30 minutes at most, depending on the film (medium-format 120 films take longer than 35 mm negatives). I use a Paterson light-tight tank (Figure 6.9). In this system, the exposed film is loaded onto a spiral which is then placed in the tank (after which point the lights can be put back on). Film spirals must be kept clean and *dry*; wet spirals are difficult, if not impossible, to load, and can result in creased films. Loading can be done in a "change bag," but I much prefer to use a properly darkened room, even though in the summer it is necessary to wait until late at night before films can be loaded. Once loaded onto a spiral, the film must be kept in the light-tight tank until processed.

Processing consists of adding *developer* to the tank for the required time, following the instructions provided with the stock chemicals. After developing, use a *stop bath* (a mild acid solution), then allow plenty of time in *fixer*; use of the stop bath extends the useful life of the fixer. Once fixed, the negative should be thoroughly washed, then taken off the spiral and hung up to dry for several hours. Resist the temptation to examine the negatives before they are properly dry – many observers have ruined good meteor photographs by touching the film while the emulsion was still "tacky."

The main points to remember when processing your own meteor films are to keep your developing tank clean, and to dry the spiral between use. If you plan to develop a lot of film – after the Perseids, say – it is a good idea to buy a few spare spirals. Photographic chemicals should, if possible, be made up using filtered water, as gritty particles from tap water can stick to the negatives. "Concertina" bottles are good

Figure 6.9 *Developing tank and spiral, ideal for processing meteor films.*

for chemical storage, as they may be gradually collapsed to exclude air – exposure to air slowly oxidizes the developer, making it less effective. Many experienced astrophotographers recommend the use of glass bottles for longer-term storage of photographic chemicals.

Do not try to use your photographic chemicals beyond the recommended number of films. I know of a couple of experienced observers who have lost valuable results through the use of old, worn-out developer. Fresh chemicals are worth the cost. Remember, you cannot repeat the exposures, and sufficient time has already been invested in photographic meteor work by the time you reach the processing stage to justify the small additional expense.

Astrophotography is a popular activity for amateur astronomers, and your local society or club may have its own darkroom facilities available for members' use. It may well be worth contacting your local group before setting out to do processing and printing.

After processing, your negatives should be carefully stored in proper labelled files, available from photographic suppliers. Provided they have been adequately fixed, your negatives will provide a permanent record of photographic meteor work for future reference.

Once dried, the negatives should of course be scanned for meteor trails, and the results reported to the appropriate coordinating body.

Scanning the negatives

For example, the BAA Meteor Section has an active photographic sub-group, from which data are collected and analysed. The pooled results of several observers allow a more accurate assessment of overall photographic meteor activity.

There are many ways of examining negatives. Rarely, the brightest meteors will leave spectacular trails which can be readily seen, even on a 35 mm negative, with the naked eye. More commonly, however, some sort of optical aid will be required to find meteor trails. A binocular microscope is excellent, if rather costly. Not surprisingly, few amateur astronomers have direct access to one, although, like other darkroom facilities, such equipment may be found in the hands of local astronomical societies. Small hand-viewers known as lupes giving a magnification of about ×8 or ×10 are commonly used, and reasonably cheap. These can be used with the background illumination of daylight or, better still, a light-box (again, your local society may have one of these). Be very careful not to scratch the negatives while examining them. Scratches can look like meteors at a casual glance. Never handle the negatives with greasy or sweaty hands.

Meteors should be apparent as long trails, cutting across the star trails in most cases. If a visual watch was carried out in parallel with the photography, you may have a good idea of where to look on the negatives. Remember, however, that the meteor trails will often appear fainter on film than they were in the sky. A general rule is that the photographic image of a meteor will appear about the same density, on film, as the trail of a star fully five magnitudes fainter. This is a consequence of the rapid motion of the meteor across the camera's field of view – there is little time for a given point along the meteor trail to build up a dense image on the emulsion.

It is worth scanning each set of negatives thoroughly at least a couple of times. Good, strong trails can, naturally, be printed. Some of the less intense images may prove rather less worthwhile, but should still be noted and reported as recorded events.

Taking it all a bit further

While the simple methods of meteor photography described above will, with perseverance, yield pleasing images and give the observer much satisfaction, anyone who has caught the meteor photography bug may well wish to look into the subject more deeply. By using similar equipment, results of some scientific value can be obtained, though getting them will require more of the observer's patience.

One obvious enhancement, as far as the assessment of rates of bright meteors is concerned, is to use several cameras. Instead of using one camera, two or more can be used, with slightly overlapping fields, to cover more of the sky at the optimal 50° altitude. Multi-camera systems can, of course, be expensive to set up, and expensive on film.

Figure 6.10 shows the battery of six Lubitels which I use during major showers. Rather than mounting each camera on a single tripod, all are supported on a common base on a table. Blocks cut so that the cameras are aimed at 50° altitude are arranged in a hexagon, each block

Figure 6.10 *The author's battery of six Lubitel 166 cameras, arranged so as to give maximal sky coverage.*

carrying a camera-mounting bush. Exposures can be made as normal, or may be synchronized by removing and replacing a matt black board over the cameras to start and end exposure for all six simultaneously. Careful records of all exposures should be made for such a system, and the observer also has to keep track of which film was exposed in which camera. I number the cameras 1–6, with camera 1 being aimed at 000° azimuth (due north), camera 2 at 060° azimuth, and so on.

While it may appear expensive, it is worth bearing in mind that the cost of putting together a battery of six Lubitels is actually no greater than for a single mid-range SLR camera. The construction of such a camera battery would make a good project for a local astronomical society – the equipment can be used to augment group visual watches.

Without any further additions, such a camera battery can be useful in a couple of ways. The advantages for determining rates of bright meteors are obvious. Lubitels have the added advantage of having a relatively long focal length (75 mm) and producing a large image. These characteristics can be usefully applied to positional work, provided the observer carefully and accurately records the start and end times of the exposure, and of the appearance of any bright meteors likely to have been recorded. For photographic work carried out in conjuction with group watches, visual observers can help greatly by aiming to note when bright meteors appear, to the nearest 3 seconds. The accuracy of exposure time can be more closely controlled: use of a matt black board is again recommended to synchronize exposure for all cameras in the battery. Some camera operators also interrupt exposures at a convenient mid-point, so that star trails are broken by small gaps (*fiducal marks*) as a further aid to positional work.

Provided sufficient care has been taken to record exposure times accurately, any resulting meteor trails can be precisely measured using a suitable microscope and later converted to right ascension and declination. For the conversion, it is necessary first to identify some of the stars whose trails are on the image, with the aid of a good star catalogue. Useful ones include *Star Catalogue 2000.0* and, more recently, computer-readable versions of the *Smithsonian Astrophysical Observatory Atlas Catalogue*. A minimal requirement in such work is for the positions of at least six field stars to be measured, along with key points (start, end, flares) along the track of the meteor. Such work is an important part of the BAA Meteor Section's photographic programme.

Positional accuracies of a few arc minutes are highly desirable. Observations of such accuracy can be used to determine the precise positions of radiants and, over longer time periods, their precise motion. For undriven cameras, the positional accuracy possible for meteor trails measured from the eventual negative depends critically on the precision with which exposure times, and the time of appearance of any photographed meteors, is recorded. Timings accurate only to the nearest minute may introduce positional errors as large as 15 arc minutes, too great to allow such undriven exposures to be used for determining radiants.

A single camera can, of course, be driven to follow the apparent motion of the stars if placed on an equatorial mount. Some telescopic equipment suppliers sell small electrically driven camera mounts designed for exactly this sort of exposure, and many amateurs use them to obtain attractive wide-field exposures of the constellations. Measurement of meteor positions relative to the star images on the resulting image is then quite simple. There are, however, obvious problems in attempting to drive a battery of cameras to follow the stars; multi-camera work is usually done undriven. However, with some ingenuity the battery could be mounted on a simple, sturdy equatorial platform.

Meteor triangulation

Pairs or multiple exposures of the same meteor photographed from locations separated by more than about 30 km (20 miles) can be used to *triangulate* the meteor's atmospheric trajectory, by the method used by visual observers since Brandes and Benzenberg in the 1790s (see Chapter 3). Photography, naturally, has the capacity for much greater positional accuracy than visual triangulation, provided the photographer takes care to record exposure times as accurately as possible.

Such possibilities further highlight the teamwork aspect of meteor observing. A single photograph is a pleasing image but, coupled with observations from elsewhere, it can become significant as a scientific record. Cooperation between local societies can lead to more work of this nature being conducted.

On a national or regional basis, the use of agreed central camera-aiming points can usefully improve the success rate for such work. Obviously, the chances of success are greatest during major showers such as the Perseids or Geminids, when bright photographic-range meteors are more abundant. It is usual to select triangulation centres to the east of a given region – most radiants lie to the eastern side of the sky when at their best for observing.

The azimuth towards which the camera at a given location should be aimed can be found quite simply from the agreed centre's bearing on a map. Camera elevation is also fairly straightforward to determine, and depends on the distance of the observing location from the triangulation centre. For triangulation purposes, cameras are aimed at a "box" of atmosphere in the meteor layer 90 km (55 miles) above the agreed centre. The elevation is therefore given by the formula arctan $90/d$, where d is the horizontal distance in kilometres from the centre.

For this work, the camera must be left in the same position for all the exposures. Again, the operator should take as much care as possible to time the start and end points of exposures.

Rotating shutters

The six-camera battery is an ideal platform for a further level of sophistication in meteor photography, the addition of a *rotating shutter* (Figure 6.11). Simply, the rotating shutter is a blade or, more usually, a series of blades spun on a central spindle by a synchronous motor, such that the sky is alternately obscured and revealed to the cameras. There are a number of advantages in using such a shutter.

Perhaps most obvious are the benefits it brings to meteor *identification*. Typically, rotating shutters should be driven to give 15–30 breaks per second. At such shutter speeds, meteor trails on film will be broken into dashed lines, as in Figure 6.12, which shows a couple of Perseids recorded by Steve Evans. Star trails remain solid, as do those of slow-moving satellites. This should ease identification of meteors when the negatives are scanned, greatly reducing the risk of misidentifying satellite trails as meteors, and improving the accuracy of any meteor rates derived. This is particularly important for multi-camera systems, as the observer will be unable to confirm visually any events recorded by cameras aimed in the opposite direction to the watch.

An additional advantage of using a rotating shutter is that it cuts down the condensation of dew on the camera lenses. The rapidly spinning blade generates air movements which make it difficult for dew to form. Use of a rotating shutter also helps to reduce the effects of sky fog, and the observer will not need to wind on the films so frequently. A 60-minute exposure made with a rotating shutter which gives an open/close ratio of 1:1 will accumulate no more sky fog than will a 30-minute exposure made without a rotating shutter.

The main scientific advantage of using a rotating shutter with cameras for multiple-station photography is that it provides images

Figure 6.11 *A camera battery, similar to the one shown in Figure 6.10, operated by Steve Evans. This battery is equipped with a rotating shutter as an aid to identifying meteor trails on the resulting negatives and to enable the velocities of photographed meteors to be calculated.*

from which the velocities of meteors can be deduced. Their atmospheric path lengths (typically 25 km, 15 miles) may be determined from photographs taken at two stations. If one station is equipped with a rotating shutter running at a known rate, the number of shutter-breaks in the meteor's trail gives an indication of its duration, from which the velocity may be determined by simple arithmetic. Rather more advanced calculations may then be applied to arrive at an approximate Solar System orbit for the incoming body. Such work remains of some value, even for well-studied showers like the Perseids.

Bright meteors occasionally leave persistent trains lasting a few seconds which are themselves sufficiently bright to register on film. There are obvious difficulties in recording these by conventional means,

Figure 6.12 *A pair of Perseid meteors photographed by Steve Evans using the rotating-shutter system shown in Figure 6.11.*

but the use of a rotating shutter can sometimes give photographic images of at least segments of such trains. Such a persistent train will have been captured when one or two of the shutter-breaks appear "filled in," as close examination of Figure 6.13 will show. The magnitude −6 Perseid, photographed by Steve Evans at 2313 UT on 1983 August 12–13, left a persistent train lasting several seconds.

Building a rotating shutter

Synchronous motors can be driven from the domestic mains (line current). However, there can be hazards in operating electrical equipment outdoors at night in damp conditions. A safer alternative has been designed by Steve Evans of the BAA Meteor Section. The Evans rotating shutter uses a bicycle dynamo as a motor, run via a low-voltage a.c. supply.

The dynamo should have a minimum working voltage of around 6 volts, and is powered from a 9-volt step-down transformer. The transformer should, of course, be protected from the damp – a sealable plastic food container may be used. It is recommended that, for further insulation, the transformer be mounted on a block of wood, and the whole unit kept indoors, with only the low-voltage feed exposed to outdoor conditions.

Loudspeaker cable delivers the reduced voltage to the dynamo, which is mounted centrally among the cameras. The shutter blade is attached under the knurled head of the dynamo (which in normal use would be in contact with the cycle wheel). Lightweight materials should be used for the blade – balsa wood or modellers' plastic card are ideal. The blade should be painted matt black to prevent any extraneous light from being reflected into the cameras. To set the blade in motion, it is usually

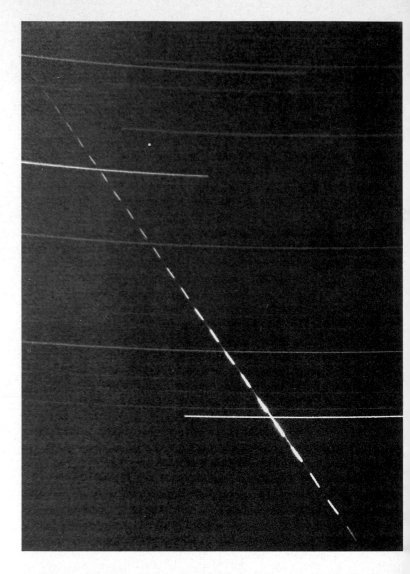

Figure 6.13 *A Perseid meteor photographed by Steve Evans on 1983 August 12–13 at 2313 UT. The meteor left a bright persistent train, which has recorded – filling in the shutter-breaks – towards the end of the trail.*

necessary to flick it gently with a finger; once spinning, the shutter will continue to do so until the power is switched off.

This design, shown in Figure 6.14, has been used by Steve Evans for over 700 hours of exposure during 11 years of operation. Fed at the UK mains frequency of 50 Hz, the dynamo spins at 12.5 revolutions per second, so a 1:1 shutter will give 25 breaks per second. For a 60 Hz line current, as in the United States, the rate of spin will be 15 revolutions per second, and a 1:1 shutter will give 30 breaks per second. Suitable blade patterns are shown in Figure 6.15.

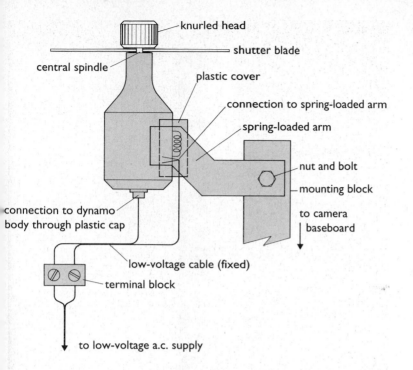

knurled head

shutter blade

central spindle

plastic cover

connection to spring-loaded arm

spring-loaded arm

nut and bolt

mounting block

connection to dynamo
body through plastic cap

to camera
baseboard

low-voltage cable (fixed)

terminal block

to low-voltage a.c. supply

Figure 6.14 *Steve Evans' design for a rotating-shutter assembly.*

The vast majority of meteor events observed to occur in the atmosphere are the result of the continual influx of small, dusty particles, none of which survive the ablation process intact (see Chapter 2). Only very rarely will even a dedicated regular observer record the arrival of a larger piece of debris, initially of the order of a metre or more in diameter, and which has some chance of being recovered on the ground as a small, solid body – a meteorite. The arrival of a meteorite is heralded by an extremely bright meteor event – a fireball.

Fireball patrol photography

Figure 6.15 *Suitable blades for a rotating shutter. Each has a 1:1 open/close ratio.*

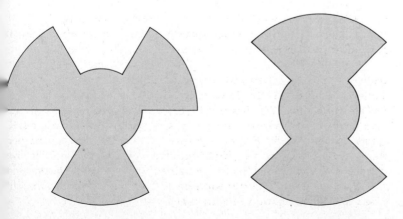

Figure 6.16 *A standard SLR camera with a fish-eye adaptor, used for all-sky fireball patrol photography.*

Visual observation of fireballs has been discussed in Chapter 4. Such events are comparatively rare, and their occurrence is far less predictable than the returns of the regular annual showers. Although there is evidence that they may be more common at certain times of the year (particularly the spring months), fireballs to a large degree occur at random, and only systematic, regular photography stands much chance of recording reasonable numbers of them. Fireball photography is conducted primarily in the hope of capturing on film the arrival of potentially meteorite-dropping events. Only extremely bright fireballs, of magnitude −10 and brighter, are likely to come into this category, so fainter objects need not be so rigorously pursued.

For several years, BAA members operated a network of 35 mm cameras, equipped with Spiratone fish-eye adaptors (Figure 6.16) for fireball patrol photography. This lens–adaptor combination was optically comparatively slow, operating at something like *f*/8. Although limited in its capacity to record moderately bright objects, the system was adequate for recording the very bright fireballs in precisely the magnitude range of interest. Anything bright enough to record with a lens system this slow had to be worth further investigation.

Typical exposures of up to 90 minutes, using ISO 400 film, were possible without appreciable sky fog. Automated systems devised by some operators allowed exposures to be made throughout every clear night. Although some success was achieved using such equipment, in that the variations in the flux of very bright events was monitored, British observers have not so far recorded an image that has led to the recovery of a meteorite. Similar patrol work in eastern Europe, Canada, and the United States has produced some rewards (see Chapter 3).

Fireball patrol photography remains a worthwhile pursuit for those prepared to undertake it. Ideally, several observers should cooperate as a team, operating all-sky cameras at stations 50 km (30 miles) apart and covering overlapping "boxes" of atmosphere. Any resulting images can, like those from conventional cameras, be used for triangulation. This sort of work could, like the operation of a multi-camera battery during national or regional projects to cover the major annual showers, be a useful area of work for a society group.

Modern photographic equipment includes 9 mm and 16 mm focal length fish-eye adaptors capable of operation at lower *f*-numbers, improving the scope for recording fainter events. Older, slower systems are not to be despised, however, and the benefit of being able to leave the camera for longer periods between frame changes is forfeited when using a faster lens system. An important limitation to be borne in mind is that fish-eye lenses are relatively large pieces of cold glass onto which dew can rapidly condense, so a lens heater is necessary. The lens should, naturally, be shielded from nearby lights – bear in mind that its wider field of view may make the all-sky system susceptible to light pollution from sources invisible to cameras used for more conventional work.

The grain of modern photographic emulsions is finer than in those in use during the 1960s and 1970s, allowing images on negatives to be measured more accurately. However, images will be quite seriously distorted, particularly towards the edge of the field of view. They often appear curved (as in Figure 6.17), unless they are short or in the very centre of the field. As for other photographic meteor work, Ilford HP5 or Kodak TriX or TMAX emulsions can be recommended.

Figure 6.17 shows a splendid sporadic fireball, recorded on 1991 September 14–15 in an exposure between 2305 and 0035 UT by Henry Soper, a regular BAA contributor from the Isle of Man. Like many others, he also enjoyed some success in photographing the 1990 Geminids. The 1h 21m exposure from 1990 December 13–14 reproduced as Figure 6.18 shows at least three Geminid trails.

The negatives should be developed and scanned carefully and systematically as for conventional meteor photography. Results should be submitted to the appropriate coordinating organization – especially if you have recorded any trails which might have been captured by photographers or seen by visual observers elsewhere. There is some risk of misidentifying aircraft or satellite trails as meteors in such exposures, but much less than for conventional photography. It is possible to equip a fish-eye system with a rotating shutter, though it will have to be placed to one side of the lens.

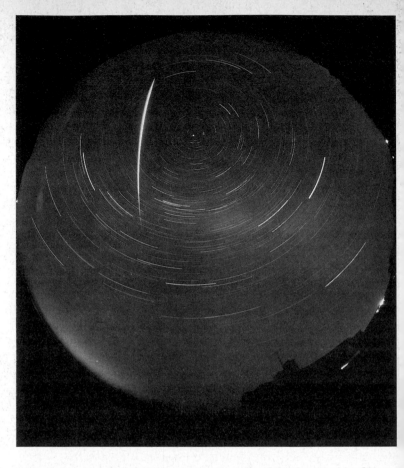

Figure 6.17 *A sporadic fireball photographed by Henry Soper from the Isle of Man on 1991 September 14–15, using an all-sky camera.*

Meteor spectroscopy

Investigation of the chemical make-up of meteoroids entering the Earth's atmosphere is difficult. The micrometeorites which eventually settle to ground level are contaminated by terrestrial material, while Brownlee particles are contaminated by the oil used to collect them. An indirect method of study is to obtain *spectra* from the brightest meteors.

Meteor spectroscopy is an area which will tax the patience of the observer yet further. While conventional photography can record meteors of magnitude +1 and brighter, spectroscopy is restricted to brighter objects, typically of magnitude −6 or better. The reason for this is simple. In conventional work, the meteor trail is focused quite sharply into a narrow strip on the emulsion, but in spectroscopy this light is spread out into a series of images whose positions depend on the wavelengths of the light emitted. Spectra can therefore be a good deal fainter, and harder to detect on the negative.

Astronomers collect stellar spectra by passing the light of the target object via a narrow slit through a prism. A similar result can be obtained by allowing the star to trail across the field perpendicular to the dispersion plane. In effect, a bright meteor also makes its own "slit" as it streaks across the field of view.

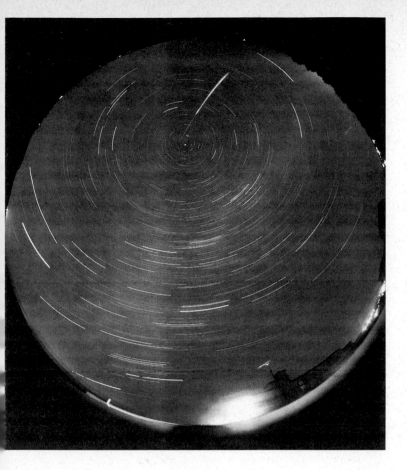

Figure 6.18 *During the 1990 Geminid display, Henry Soper succeeded in capturing several bright meteors using his all-sky camera system. At least three are visible in this 1h 21m exposure from the night of 1990 December 13–14. The brightest is probably at least magnitude −6.*

Meteor spectra are obtained by placing a prism in front of the camera lens (Figure 6.19), and taking time exposures. Prisms are preferred to diffraction gratings, since gratings break up the light into spectra of several orders, making even the brightest (first-order) spectrum rather faint.

In the system used by Harold Ridley and Steve Evans to obtain the superb Perseid spectrum shown in Figure 6.20, a 30° flint prism sits in front of the camera lens. A plate camera is used so as to give as large an image as possible. It would also be possible to use a 35 mm or – perhaps better – a large-format camera. Spectra for the very brightest meteor *trains* may, very rarely, be obtained using equipment complemented by a rotating shutter. Fast films and camera lenses are recommended. The spectrum shown in Figure 6.20 was recorded on "5 × 4" HP5 film, and was initially believed to be the spectrum of the meteor photographed in 1991 by Michael Maunder which appears on the front cover of this book; further investigations suggested that it may be of a different Perseid fireball that appeared around the same time. Just how bright the

Figure 6.19 *A meteor spectrograph: a prism is placed in front of the camera lens.*

Figure 6.20 *A highly detailed Perseid spectrum photographed at 2346 UT on 1991 August 12–13 by Harold Ridley and Steve Evans. This intensely bright meteor was seen through patchy cloud, which accounts for the light-spill in the image.*

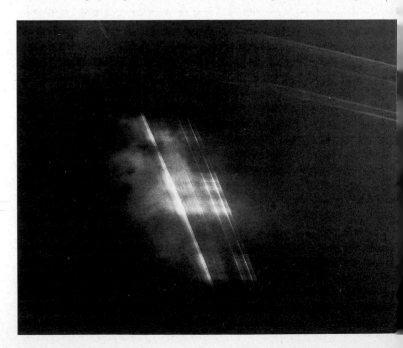

Figure 6.21 *In addition to recording its trail on his all-sky camera, Henry Soper also succeeded in recording an extremely detailed spectrum for the meteor shown in Figure 6.17 using a diffraction grating. Over 70 lines are present in the brightest, first-order spectrum. Measurement of these lines reveals a great deal about the chemical composition of the incoming meteoroid.*

fireball was is clear from the fact that the image was recorded through patchy cloud (which accounts for the "spilled light" around the flares).

Later the same year, on the night of September 14–15, Soper was rewarded for his many hundreds of hours of effort with a rarely captured, highly detailed sporadic meteor spectrum, shown in Figure 6.21. Over 70 lines are visible in the brightest, first-order spectrum, making it the most detailed recorded from northwest Europe in many years. The meteor itself (shown in Figure 6.17) is visible at the centre of the image, surrounded on either side by first- and second-order spectra. The spectra on the left are brighter, since the grating Soper used was "blazed" to shed more light to one side. Star images and their spectra (at right angles to the meteor spectrum) are also present.

Spectral images such as these are measured under a microscope to obtain precise wavelengths for the various emissions. It is usual to take three prominent common emission lines as standards against which to measure the others. High-velocity meteors usually show the readily identifiable emission lines of sodium (a doublet), magnesium (a merged triplet), and the calcium H and K lines. Slower meteors show an iron line (Fe I).

The spread of the emission line images of a meteor – the *dispersion* – depends on the angle between the meteor's apparent path in the sky and the prism's plane of dispersion. Evans and Ridley give the formula $D_{eff} = D_{max} \times \sin\theta$ for the *effective dispersion*, where D_{max} is the maximum possible, and θ the angle between the plane of dispersion and the meteor's path. The best results are obtained when the meteor is perpendicular to the plane of dispersion, a rare occurrence. Spectra are more likely to be well spread-out if the prism is oriented so that any meteors emerging from the shower radiant under study will have apparent paths close to the optimum. Sporadic meteor spectra are, not surprisingly, extremely rare, and difficult to obtain.

Video recording of meteors

Modern technologies are quickly embraced by amateur astronomers in a number of fields. Photoelectric photometry for the measurement of variable-star magnitudes is one example, the use of charge-coupled devices (CCDs) for deep-sky imaging is another. Low-light cameras, based on CCD technology, have been used commercially for security purposes in factories and other areas left unattended at night. Obtained second-hand, such cameras have been turned to astronomical use by a number of resourceful amateurs. Low-light video imaging has been successfully used to record meteors. Although this method is now still in its infancy, at least among amateur astronomers, it has the potential to become a valuable means of augmenting existing photographic techniques.

One way in which this work has been carried out, by Andrew Elliott, an amateur astronomer based at Reading, England, has been to fit a 28 mm wide-angle lens in front of the low-light camera, the image from which is recorded with a video camera. The resulting tape can be viewed at leisure, and any meteors recorded noted for reference. Some of the results have been spectacular – particularly the records of slow, bright Geminid fireballs in 1990.

A great advantage of such a system is the possibility of improving the accuracy of timing. A digitized broadcast time signal can be fed into the video, allowing a timing precision of 0.04 second for events on screen (for a frame rate of 25 frames per second, 0.033 second for the US rate of 30 fps). This can be used to enhance the precision of positional measurements taken from conventional photographic negatives exposed while the low-light/video system has been in operation.

Video recording of meteors also has its aesthetic appeal. It is pleasing to review Perseid or Geminid activity at a later date, and see again – in real time – bright meteors or short bursts of several meteors in quick succession. As a way of introducing beginners to the idea of a meteor radiant, and how meteors appear in the sky, video has educational potential. More research is required to determine how useful video will prove in the assessment of meteor activity and other analyses.

CHAPTER 7

Other methods of observing meteors

While most observers are content with naked-eye watches, perhaps augmented by photography, it is also possible to record meteor activity by telescopic means, or by radio. Both these alternatives require some outlay either financially or in terms of the observer's patience, but there are rewards for those who persevere.

Most "telescopic" meteor observations are, in fact, carried out using binoculars. Visual observation, as outlined in Chapter 4, exploits the wide field of view of the human eye. Binocular observing makes use of the superior light-gathering power of the optical system to bring into view fainter meteors, down to about ninth magnitude. The narrower field of view, however, results in fewer meteors being seen by the binocular observer. In many respects, telescopic meteor work is a discipline to itself, requiring considerable patience. In particular, it takes a dedicated observer to forego the excitement of observing the naked-eye Perseid peak in order to concentrate on recording the fainter shower members.

Telescopic observing

Equipment

Binoculars are the instrument of choice for most telescopic meteor work. They are described in terms of their magnification and aperture. The types most widely used by amateur astronomers are binoculars of 8 × 30, 7 × 50, or 10 × 50 specification. In each of these the first figure is the magnification (8, 7, and 10 × respectively), and the second is the aperture (diameter), in millimetres, of each of the objective lenses. Many amateur astronomers use 10 × 50 binoculars for variable-star observing, or to obtain wide-field views of deep-sky objects. Financially, such binoculars are probably the best investment for the beginner. A pair of 10 × 50s gives a field of view approximately 5° across, quite adequate for telescopic meteor work.

As for naked-eye visual work, the observer's comfort is extremely important. It becomes quite a strain to hold a pair of binoculars steady while looking up at the sky for any length of time. The best solution, again, is for the observer to recline in a deckchair, with the binoculars comfortably accessible from a sitting position. The binoculars must be mounted on a tripod, since you will need your hands free for recording.

Photographic shops can usually supply, at a reasonable price, suitable small adaptors which allow binoculars to be clamped to a camera tripod. Special binocular holders are available in the United States.

Small "rich-field" Schmidt–Cassegrain or catadioptric telescopes have been used for such work, particularly in professionally coordinated observing programmes in eastern Europe. It is best to use a low magnification with such instruments, both to provide a reasonably large field, and for "eye relief" – high-magnification eyepieces cause eye-strain when used continuously for long periods, as is necessary for telescopic meteor work.

Observing routine

In principle, telescopic meteor work is little different from naked-eye watches. The smaller field of view and lower capture rate, however, place some constraints on telescopic work, and a different observing routine has to be adopted.

The low rates will obviously have an effect on the observer's attention span. It is better in telescopic work to do three 20-minute stints, separated by reasonable rest periods, than to slog on for a subjectively very tiring unbroken hour. It is also more tiring on the eyes to have a relatively limited field of view. The observer will be in better shape to actually see the one or two telescopic events which occur in the field if breaks are taken.

Telescopic watchers usually select a couple of fields, one either side of a known radiant. A good distance is probably about 20° in azimuth from the radiant, at the usual elevation of about 50°. Malcolm Currie, a very experienced telescopic meteor observer, advises against using field centres closer than 10° to the radiant.

Field selection is quite important. Fields containing bright stars should be avoided – objects such as Capella or Deneb will be obvious distractions, and their glare may cause the observer to miss faint meteors. Reasonably rich backgrounds are good as they provide plenty of stars against which the paths of meteors can be defined in subsequent plots.

Charts showing selected fields are prepared by national organizations such as the BAA Meteor Section for use during specific projects. If all observers enagaged in a particular project are using the same fields, it becomes easier to collate their results. Solo observers working at other times may consider preparing their own field charts in advance, by tracing relevant sections from an appropriate star atlas; *Uranometria 2000.0*, which shows stars down to about ninth magnitude (about the same limit as 10 × 50 binoculars in a good sky) is perhaps the most useful.

Your eventual watch record should contain details of the chosen field's centre in right ascension and declination. This, again, will be standardized on project sheets, or can be readily found from an atlas. Other details to be recorded include sky conditions during the watch: an indication of the faintest star visible in the binocular or telescopic field will define the limiting magnitude. Passing clouds, moonlight, or other external conditions should be noted; if cloud begins to interfere, it is probably best to abandon the watch. The *start and end times* of the watch, in UT, must always be given, along with the *duration*. *Equipment* used must also be noted.

During the watch itself, the observer concentrates on the chosen field for 20 minutes, recording details of any meteors seen. In telescopic work there is still some value in *plotting* observed meteor paths against the background stars. The more restricted field of view is actually an advantage in this work, and telescopic meteor plots can be of a very high accuracy. Such plots have been used to refine radiant positions and structure.

Each meteor plot should only be made on the field chart once the observer is absolutely certain of the trajectory relative to the background stars. Each plot should include a small arrow, indicating the *direction* of motion, and each meteor should be *numbered* for later identification so that other details may be noted. These other details include *time* of appearance (UT) to the nearest minute, a rough visual *magnitude* estimate, and marked *colour*, flaring, or *persistent train* phenomena.

During a telescopic meteor watch, meteors will be seen to enter or leave the field of view, or appear in their entirety in the field. The type of sighting should also be noted on the report: *aa* (in some standards 11) denotes a meteor appearing entirely within the field; *ao (10)* one that starts in the field, then leaves; *oa (01)* one that starts outside the field, then enters.

Figure 7.1 shows an example of a telescopic watch report. A typical binocular watch will yield only 2 or 3 meteors per hour, and I must stress again that dedication and patience are essential qualities for observers wishing to attempt telescopic meteor work. Although observation of the fainter meteors visible in telescopes can reveal fine structure in known shower radiants, the overwhelming majority of very faint meteors are actually sporadic in origin.

After 20 minutes the observer should stop the watch and rest for at least 10 minutes. The watch resumes, concentrating on the second field, on the other side of the radiant from the first. Watches should continue to alternate from one field to the other for as long as observations continue on a given night. Again, it must be stressed that it is quality rather than quantity that is of prime importance in telescopic meteor work. It is better to do a couple of reliable watches on a given night than to struggle on for long periods, recording poor-quality data in extended sessions.

Binocular observation of persistent trains

Rarely, bright naked-eye visual meteors may leave behind them ionization trains persisting for several minutes. Binoculars are ideal for observing these in detail, and might be considered a useful additional item in the visual observer's equipment. Binocular observation of persistent trains requires fast reactions. Most trains will fade from view in the space of only a few seconds at most. Longer-duration events should be followed in binoculars as they slowly fade from view. Rough sketches can be made, showing the gradual evolution of such trains.

Winds in the high atmosphere cause trains to drift and distort, sometimes quite rapidly. Initially straight trains have been observed to form into small loops as a result of high-atmosphere wind-shear over

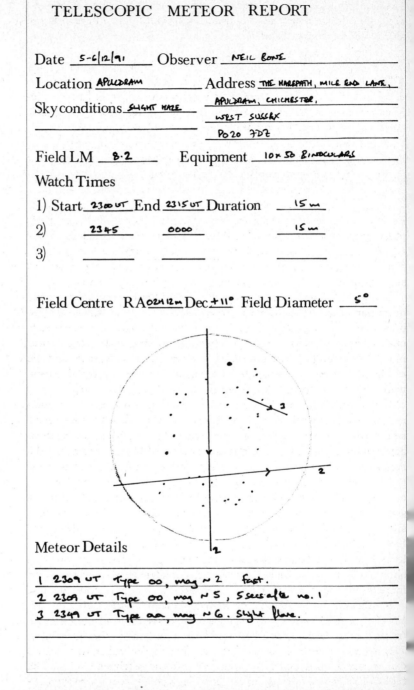

TELESCOPIC METEOR REPORT

Date _5-6/12/91_ Observer _NEIL BONE_

Location _APULDRAM_ Address _THE HAREPATH, MILE END LANE,_

Sky conditions _SLIGHT HAZE_ _APULDRAM, CHICHESTER,_

WEST SUSSEX

_PO20 7DZ_

Field LM _8.2_ Equipment _10 x 50 BINOCULARS_

Watch Times

1) Start _2300 UT_ End _2315 UT_ Duration _15 m_

2) _2345_ _0000_ _15 m_

3) _____ _____ _____

Field Centre RA _02ʰ12ᵐ_ Dec _+11°_ Field Diameter _5°_

Meteor Details

1 2309 UT Type oo, mag ~2 fast.

2 2309 UT Type oo, mag ~5, 5 secs after no. 1

3 2349 UT Type aa, mag ~6. Slight flare.

Figure 7.1 *A telescopic meteor report. Observed rates are lower than for naked-eye meteors, but fairly precise positional measurements can be made and it is this that gives the work its interest and value.*

the course of a couple of minutes. Radar observations of meteor trails and photographic triangulation of noctilucent clouds (see Chapter 8) both point to wind speeds of as much as 400 km/h (250 mile/h) at the altitude of the meteor layer.

For the few moments of its existence, a meteor's ionization trail can act as a "surface" from which radio waves are reflected (or, in radio parlance, *scattered*). Professional astronomers began to use radar and radio techniques to detect meteors in the 1940s. As described in Chapter 3, much of the pioneering work was carried out from Jodrell Bank in England. The first studies used long radio wavelengths, but it was soon found that short wavelengths were more useful.

Radio observing

It was determined that an ion density of 10^6 per cm^3 was sufficient to give clear reflections of signals from meteor events. Two regions of ionization were found to be associated with meteors – the trail itself and, preceding it, a denser region at the head. Meteor reflections may be either *underdense* (sharply peaking in intensity, and falling away rapidly and exponentially), or *overdense* (persisting at maximum intensity for several seconds). Overdense reflections are produced by larger meteoroids.

Bernard Lovell and his colleagues at Jodrell Bank used the technique of radar reflection (*backscatter*). The delay between sending out a radio pulse and detecting its reflection back at the source was used as a means of determining meteor velocities. The intensity of reflection for shower meteors was found to depend on the angle between the radiant and the receiver; meteors appearing 90° from the radiant gave strong reflections, while those close to the radiant were not usually detected.

It is possible for modern amateur astronomers to carry out radio meteor studies with less sophisticated equipment. Amateurs have restricted access to radar equipment. Around the world, however, there are thriving bodies such as the Radio Society of Great Britain and the American Radio Relay League which promote and coordinate amateur radio as a pastime. Most radio "hams" are interested primarily in communicating with fellow enthusiasts over long distances, by reflecting signals from aurorae, for example. Some radio amateurs have also shown an interest in using meteors to reflect signals over long distances. Under normal circumstances, amateur radio transmitters, as used in the legally permitted 4-metre band in the UK, have a maximum range of about 160 km (100 miles). Reflection from meteors or aurorae can increase the range of signal.

In amateur meteor detection, the technique of *forward scatter* is used. Forward scatter enables VHF signals to be detected from a transmitter (radio or television) which lies over the normal radio horizon (Figure 7.2). The brief ionization trail of a meteor in the upper atmosphere can scatter the transmitter's signals towards the receiving station. The rate of occurrence of such reflections – "pings," in radio ham jargon – gives an indication of the meteor flux, and varies according to the time of day and the presence of shower activity. Pings can be recorded automatically on a chart recorder, or may be counted manually in real

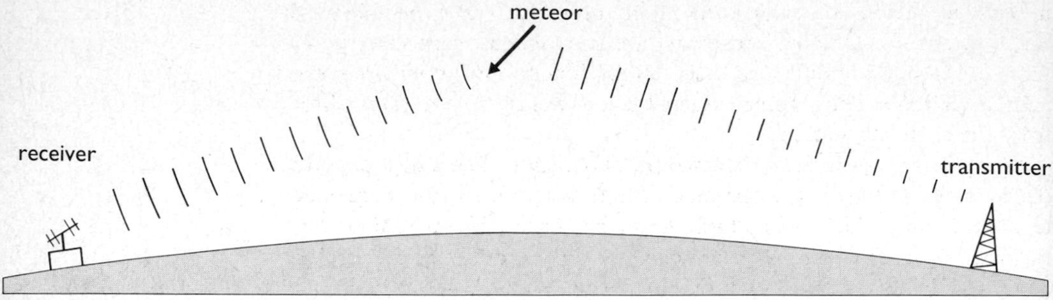

Figure 7.2 *The principle of forward-scatter radio meteor detection. Signals from a transmitting station lying beyond the receiver's normal horizon can be detected when they are reflected from meteors lying between the transmitter and receiver.*

time by the operator. Reflections are detected as brief snatches of music or speech above the background noise level.

For this type of work, a range of about 1000–1500 km (600–900 miles) between transmitter and receiver is ideal. Meteor trails will be detected in the atmosphere roughly 100 km (60 miles) above the mid-point of the transmitter–receiver path. The receiver, which is fed from a four-element Yagi antenna, may be tuned to distant stations broadcasting on the 4-metre band from 66 to 70 MHz. Several such stations in eastern Europe are ideally located for UK operators; observers in the United States can use FM stations lying a suitable distance away. It is best to choose a station which broadcasts for a full 24 hours each day. Details of suitable transmitting stations may be obtained from the annual *World Radio TV Handbook*.

The duration of a radio meteor's ping depends on the size of the incoming meteoroid. Most are small, and give rise to very short events, but larger meteoroids – such as those which produce visual meteors of around second magnitude – can cause pings which persist for several seconds. The sensitivity of the receiver can be adjusted so that only the more substantial meteors are recorded. The sensitivity can be lowered by tuning to lower-power transmitting stations, thereby recording only those pings which result from larger meteoroids.

Suitably equipped operators in the UK may also detect forward scatter from Scandinavian or eastern European television stations. John Branegan operated such a system from Fife, in Scotland, in the early 1980s, detecting bursts of signal from television stations in southern Sweden in the 6-metre band. He made counts of the number of reflections in 5-minute intervals. More powerful reflections occasionally resulted in a "lock," in which a complete television picture was received. More usually, meteor reflections produced an audible crackle, and a fragmentary picture.

The forward-scatter technique does not allow specific identification of meteors – it cannot distinguish between Perseids and sporadics, for example. Given a knowledge of sporadic activity throughout the year, however, the background can be subtracted from the overall total to give some idea of shower rates. Account must also be taken of the shower radiant's position relative to the transmitter–receiver path, the *observability function*. The observability function is lowest when the azimuth of the radiant is more or less the same as that of the transmitter from the receiving station.

Sporadic meteor rates as detected by VHF radio reach a peak around dawn. When receivers are operated at high sensitivities, the frequency of pings is much higher – so high that it becomes difficult to filter out the shower activity from the already high background levels. The high activity of the early morning hours has been affectionately termed the "dawn chorus."

John Branegan's work on television reception yielded similar results, with the reflection rate peaking in the early morning hours. Around dawn, 10–15 strong television signals might be received per hour under non-shower conditions using a receiver set to a relatively low sensitivity. During the peak of the Perseids in 1980, the level went up to 60 reflections per hour.

Radio scatter can detect meteors well below the naked-eye visual range, theoretically down to a magnitude of +15. It is therefore difficult to use radio methods to derive equivalent visual rates. Radio observation does have the advantage under cloudy conditions, and comes into its own when a short-lived shower peak occurs during daylight.

Radio work is not free of natural problems. There is the as-yet poorly understood phenomenon of *sporadic E*, which occurs during the summer and can result in the loss of observing time, while auroral activity can also disrupt meteor detection. Sporadic E manifests itself as thin sheets of ionization in the atmosphere at altitudes of around 100 km (60 miles), the E-layer of the ionosphere, which may persist for several hours. Reflection from Sporadic E leads to interludes of continuous reception from distant stations; conventional television reception is also affected to some degree. Radio meteor detection becomes impossible under such conditions. A prolonged spell of Sporadic E prevented UK radio observers from following up their successful coverage of the 1980 Perseids in 1981.

The main interest of radio amateurs in meteor forward scatter is as a means of communication. Messages can be compressed into short bursts for continuous transmission at the widely used 144 MHz frequency. An operator receiving such a message via meteor scatter can later decompress it and read it. However, a small handful of amateur astronomers have taken a specific interest in radio work, principally with a view to monitoring meteor activity. The equipment can be expensive, but it is also possible to pick up useful receivers at low prices second-hand: it appears that, like hi-fi buffs, many radio hams are not satisfied unless their system is the very latest model! It is worth watching the small advertisements in the pages of enthusiasts' monthly magazines such as *Practical Wireless* in the UK, or the American Radio Relay League's *QST*, for suitable equipment on offer. Local radio clubs should also be a valuable source of advice, and are well worth contacting.

Care should be taken to shield the receiver and its amplifier from domestic interference, which comes from a diverse range of sources including televisions, electric lawnmowers and vacuum cleaners. The domestic mains supply (line current) is also variable, and a smoothing unit must be used to maintain a steady filtered voltage feed to the radio equipment. Once set up, the system is capable of yielding results of scientific value, but the would-be operator has to be prepared to part with a fair amount of money, and be very patient.

John Mason used his equipment at Barnham in Sussex to monitor the Giacobinids for enhanced activity in 1985, as the Earth passed the descending node of the stream's orbit 26 days after the parent comet, P/Giacobini–Zinner (see Chapter 5). His results, indicating a significant outburst of Giacobinid activity in the early daylit hours of 1985 October 8 (also seen visually by Japanese observers), provided a well-deserved reward for a considerable amount of time spent getting the system operational (Figure 7.3). The peak radio Giacobinid rates were detected at 0935 UT (±1 minute), 3h 40m earlier than the time of the Earth's passage through the descending node of the comet's orbit. Observations such as this allow the distribution of meteoroids near the comet's nucleus to be determined.

As with other high-cost ventures, the construction and operation of a radio receiver for meteor forward-scatter detection could, perhaps, be tackled as a group project by local astronomical societies seeking useful activities for those all-too-frequent cloudy nights. A number of societies in England have undertaken such projects, with a fair degree of success.

Figure 7.3 *John Mason's radio observations of the 1985 Giacobinid outburst, showing the sharp peak in radio reflection at 0935 UT on October 8.*

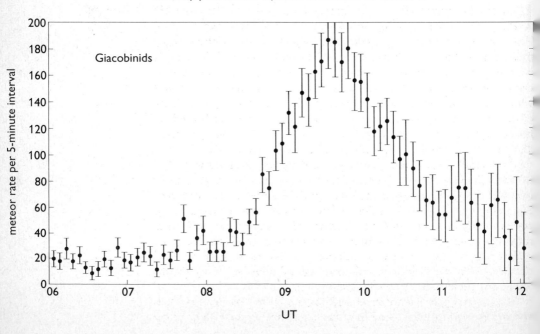

CHAPTER 8

Other phenomena

The dedicated visual meteor observer, who spends long hours outdoors under the clear night sky, can soon become familiar with the constellations, and with the movements of the planets against the background stars. It was their familiarity with the constellations that enabled William Denning to discover Nova Cygni in 1920, and J.P.M. Prentice to discover Nova Herculis in 1934. Another meteor observer, the Scottish amateur astronomer Thomas Anderson, discovered Nova Persei in 1901. From time to time other astronomical phenomena, some of them more directly related to meteors, or happening in the same general region of the atmosphere, may be seen during a watch.

Particularly for large-scale phenomena such as aurorae, the meteor watcher has an advantage over his or her dome-bound colleagues. For me, one of the pleasures of meteor observing is being out under the open sky, and seeing it as a whole, rather than viewing only a small part of it through a narrow opening in an observatory dome.

Comets

Few comets reach naked-eye brightness, and fewer still produce the long, spectacular tails often portrayed in popular astronomy books. The average comet is a faint telescopic object, only well observed by those who specialize in such work. While professional patrols, using wide-field Schmidt cameras, account for many modern comet discoveries, between a third and a half are still made by amateur astronomers who conduct dedicated visual searches. An excellent working knowledge of the sky's appearance in binoculars, built up over many hundreds of hours' sweeping, is usually required for such searches to be successful, and comet hunting is not an activity to be undertaken lightly.

Many newly discovered comets are rather faint (magnitude +10 or below) and difficult to see. Some of the amateur discoveries, and one or two of the professional finds, do brighten to become binocular objects. Such comets can generate a fair degree of excitement. Once the new comet's motion against the sky background has been followed accurately for a couple of days, its future position can be predicted. Detailed *ephemerides* of the comet's coordinates are issued in early-warning circulars from the major astronomical organizations, and may also be found on specialist electronic bulletin boards.

Similar ephemerides are produced for the known periodic comets, and are published some time in advance of their expected perihelion return. Unfortunately many periodic comets are rather dim, even at best, and will not easily be located by the more casual observer. Comet P/Halley was an obvious exception in 1985–6, being an easy binocular object for several months before perihelion. Comet P/Encke, the parent of the Taurids, is rather more difficult, usually reaching only eighth or ninth magnitude around perihelion.

Other periodic comets are more difficult to predict. Gas ejection from the nucleus can slow down or speed up a comet's motion, such that it arrives back at perihelion later or earlier than expected. Comet P/Brorsen–Metcalf returned to perihelion 18 months late in 1989, for example. Uncertainties in a comet's precise orbit also present problems, as shown by the failure of the Perseid parent, Comet P/Swift–Tuttle, to appear as expected in the early 1980s.

The chief uncertainty with comets, however, lies in predicting how bright they will be. Perhaps best known in this respect is Comet Kohoutek, which came to perihelion in 1973–4. Detected while still a great distance from the Sun, it was expected to be very bright, reaching as much as magnitude -4, once in the inner Solar System. However, when it arrived, Kohoutek was only a rather modest fifth-magnitude comet – brighter than most, but still a disappointment for an expectant public. More recently, Comet Austin in 1990 failed to live up to expectations. Predicted comet magnitudes are only a guide, and the object may well turn out to be brighter or fainter than expected.

Quoted comet magnitudes are *integrated magnitudes*, which measure the overall brightness of the object. Deep-sky observers will be familiar with the manner in which integrated magnitudes can give the misleading impression that an extended object should be bright. A comet stated to be of magnitude 6 may have its light spread over a relatively wide area, giving it a very low surface brightness, and little contrast against the background sky. Since they are brightest only when close to the Sun, comets are usually to be seen at their best low in the early evening or pre-dawn twilight. It is therefore often necessary to observe them in a brighter sky than would be ideal, which can exacerbate the problem of low surface brightness.

Perhaps a couple of comets will attain reasonable, binocular magnitudes each year. Meteor observers may wish to follow such objects, bearing in mind the fundamental role of comets in the production of meteor streams. New comets, however, are themselves unlikely to be a source of meteor activity in the short term.

Observing comets

Skilled observers can of course detect faint comets, both visually and photographically, using large telescopes. The brighter objects can be detected using binoculars, and it is these which will obviously be the favoured targets for more casual observers.

The same 10 × 50 binoculars used for telescopic meteor work are ideal for observing comets; larger-aperture (say 15 × 80) binoculars may give a further advantage. To locate comets, a good star atlas such as *Uranometria 2000.0* is essential; from the ephemeris, the comet's position

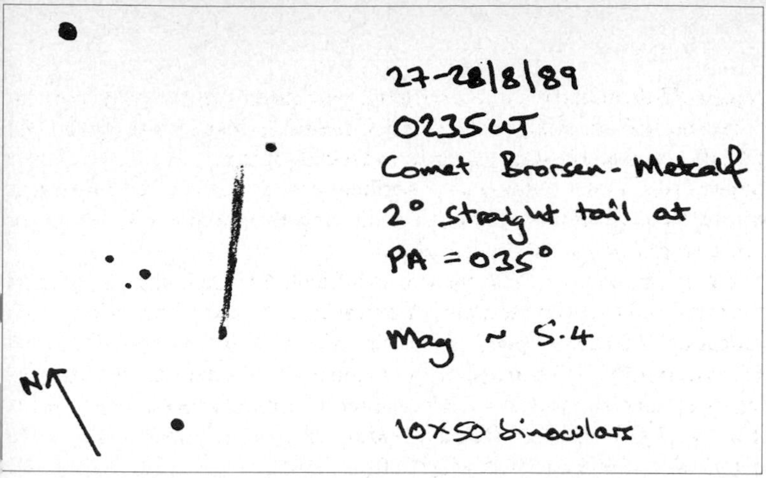

Figure 8.1 *A rough sketch of the reasonably bright comet P/Brorsen–Metcalf, and nearby field stars, made in 1989 August .*

can be plotted on a copy of the relevant chart. To find the comet, it may be necessary to "star-hop" from the brightest field star via stars progressively closer to the expected position. If bright and condensed, the comet may be easily identifiable as a small fuzzy object. The chart should of course be checked for any deep-sky objects (some of which – as Charles Messier well knew – appear similarly small and fuzzy!) which might be mistaken for the comet. More diffuse, fainter comets may require the use of averted vision: look slightly to one side of where the comet should be (the most sensitive part of the retina lies just outside its centre). It may take some time, and not a little patience, to find your target.

Once located, the comet may be examined in more detail. An obvious next step is to make a sketch of its appearance, including field stars for scale and orientation. Figure 8.1, is a rough sketch I made of Comet P/Brorsen–Metcalf on 1989 August 27–28 at 0235 UT. Averted vision may reveal the presence of a tail, or tails, the extent of which may be estimated in terms of the binocular field of view, or from subsequent measurement of sketches relative to the known scale of the chart. The *position angle* of a tail (from north, 000°, through east, 090°, and so on) may be determined from such sketches. The diameter of the coma at the head of the comet may also be estimated. Observers usually give some indication of the *degree of condensation* of the coma, which, as noted earlier, may range from compact to extended.

Magnitude estimates are difficult to obtain accurately. One widely used method is to defocus the binoculars until field stars have a similar, diffuse appearance to the comet. The observer than has to choose two field stars – one brighter than the comet, and one fainter – and decide, from memory, where the comet lies in terms of relative brightness between them. Simple arithmetic will yield a numerical value for the comet's magnitude.

Observations on successive nights will usually reveal a comet's motion relative to the background stars. The same star-hopping technique can be used to locate asteroids, and follow their night-to-night movements.

A surprising number of these objects come within the range of 10 × 50 binoculars. It should be easy enough to find Vesta, which can reach sixth magnitude (making it theoretically a naked-eye object) at favourable oppositions, while other prominent asteroids – in particular Ceres, Pallas, and Juno – can also be found, given patience and a knowledge of their position. Magnitude estimates of these minor planets can be made relative to (focused!) background stars: small fluctuations over the course of a few hours may reveal asteroids' rotation periods.

An interesting challenge is to record asteroids on film. I have succeeded in capturing the motion of Vesta on short (10-second) undriven exposures on successive nights with a 50 mm $f/1.8$ lens and ISO 400 film.

Regular comet observers may use more sophisticated techniques to estimate magnitudes and record other details; these guidelines are included here simply to allow meteor observers to pursue and appreciate the brighter objects.

The zodiacal light and the *gegenschein*

The eventual end point in the evolution of a meteor stream, as described in Chapter 1, is for it to become so dispersed that its meteoroids are no longer recognizable as having had a common source. The stream joins the general dusty background which fills the inner Solar System, and gives rise to the elusive sky glows of the *zodiacal light* and *gegenschein*.

Sadly, few present-day amateur astronomers have seen either of these phenomena, thanks to the ever-increasing spread of artificial light pollution. The zodiacal light (sometimes referred to as the "false dawn") is usually faint, comparable in brightness to the Milky Way. From the temperate latitudes where most amateur astronomers live, the zodiacal light is never particularly high in the sky, and its visibility depends on the time of year. Since it is in the plane of the ecliptic, the zodiacal light is best seen when the ecliptic makes a steep angle with the horizon, during spring evenings or autumn mornings, as shown in Figure 8.2. Observers at lower latitudes, where the ecliptic makes a steep angle with the horizon all the year round, have a better opportunity to see the zodiacal light.

Figure 8.2 *The visibility of the zodiacal light from temperate latitudes, best seen on autumn mornings or spring evenings from dark-sky sites.*

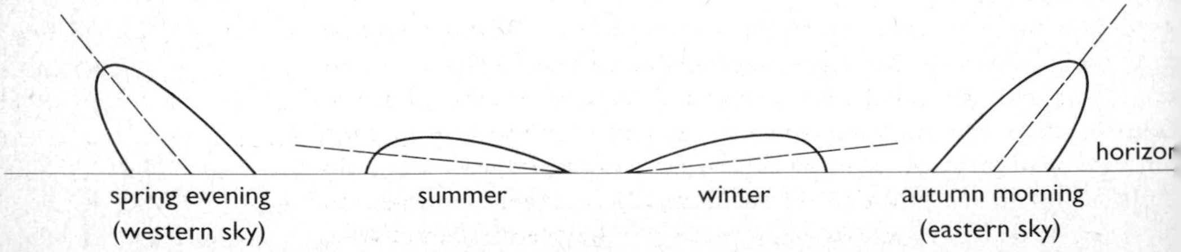

spring evening (western sky)　　　summer　　　winter　　　autumn morning (eastern sky)　　　horizon

— — — — — — — plane of the ecliptic

The zodiacal light is brightest in the direction of the Sun, and is visible for only a couple of hours after sunset, or before sunrise, at best. There have been suggestions that the brightness of the zodiacal light may change with time, varying with solar activity. It appears that the zodiacal light is brighter around sunspot minimum, when the solar wind flowing out from the Sun through interplanetary space is dominated by steady particle streams emerging from weak regions in the solar magnetic field. Excitation of the tenuous gaseous interplanetary medium by energetic particles may play a role in increasing the brightness of the zodiacal light at these times.

To observe the zodiacal light, you will need a clear, dark eastern horizon in early September, or western horizon in early March. Look for a diffuse cone of light extending upwards from the horizon along the ecliptic. The ecliptic is shown in *Norton's* or other good amateur star atlases. In September, the zodiacal light's faint glow may be seen extending as far as Castor and Pollux in Gemini a couple of hours before sunrise, while in March it may reach past the Hyades in Taurus. It might be worth making rough sketches of its apparent visual extent.

Photography of the zodiacal light is, naturally, difficult. The levels of residual twilight at those rare times when the zodiacal light is just about noticeable to the naked eye may be sufficient to swamp it on photographs. At lower latitudes, however, observers have succeeded in recording the zodiacal light using exposures of 5–10 minutes at $f/2$ on ISO 400 film. A wide-angle lens probably gives some advantage, but beware of being misled into thinking you have recorded the zodiacal light by the uneven illumination – *vignetting* – at the edge of the field.

Even more elusive is the *gegenschein*, or counterglow, which is produced by the reflection of sunlight from particles directly opposite the Sun in the sky. The *gegenschein* appears as an oval glow about 10–20° in diameter, and is weaker even than the zodiacal light. In theory, it should be best seen on winter nights when high in the midnight sky among the stars of Gemini. It is likely, however, that at this time it becomes indistinguishable from the weak light of the winter Milky Way, which traverses this part of the sky. Perhaps the best months to look for the *gegenschein* are November and February.

Extremely dark skies are required to reveal the *gegenschein*, and few astronomers can claim to have seen it in the past thirty or forty years; light pollution may well have robbed us of the chance of ever seeing it again. It is rather like trying to find a diffuse, low-surface-brightness comet in strong twilight.

Meteor streams may be regarded as denser regions within the zodiacal dust cloud. There is the interesting possibility that, under rare and special circumstances, meteor swarms, such as those of the Leonids or Giacobinids, might be observable as they approach the Earth. Some prior knowledge of an approaching swarm's expected position would obviously be required before one could attempt to record it. I suspect, however, that, like the *gegenschein*, it is likely to be overwhelmed by the general level of background nighttime illumination in the skies over most places on Earth today.

Artificial satellites

Observers at higher latitudes, such as those of Canada or the British Isles, will frequently see artificial satellites pass across the sky during meteor watches at times when the Sun is not too far below the horizon. At such times – on July and August nights, for example – satellites remain in sunlight above the Earth's shadow at typical orbital heights of 200–300 km (120–200 miles). On early spring evenings and short summer nights, satellites will often appear to outnumber meteors.

Satellites appear as points of light moving slowly against the background stars. Observers who specialize in observing them make precise timings (accurate to 0.1 second, where possible) of the arrival of satellites at specific points in the sky, such as the line between two stars or the apex of a right-angled triangle formed by two stars and the satellite, as shown in Figure 8.3. Such observations yield information on how satellite orbits are modified by changing atmospheric conditions (the expansion and contraction described in Chapter 2), or by small local fluctuations in the Earth's gravitational field.

Satellites often vary in brightness as they rotate or tumble. A regular meteor observer may come to recognize satellites appearing on successive nights on the basis of their characteristic fluctuations in brightness. Changes are sometimes slow, producing on film the meteor-like trails which are the bane of meteor photographers. Satellites are often seen to be eclipsed by, or to emerge from, the Earth's shadow.

Meteor observers, and others who spend long hours outdoors under clear, dark skies, may on rare occasions be fortunate to witness a satellite re-entry, which appears as a relatively slow-moving, long-duration fireball, as discussed in Chapter 4.

Figure 8.3 *Satellite observers make timings of when satellites arrive at certain positions (×) on the celestial sphere relative to the background stars.*

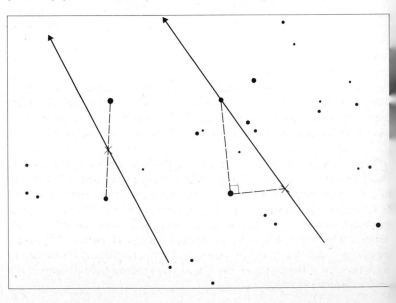

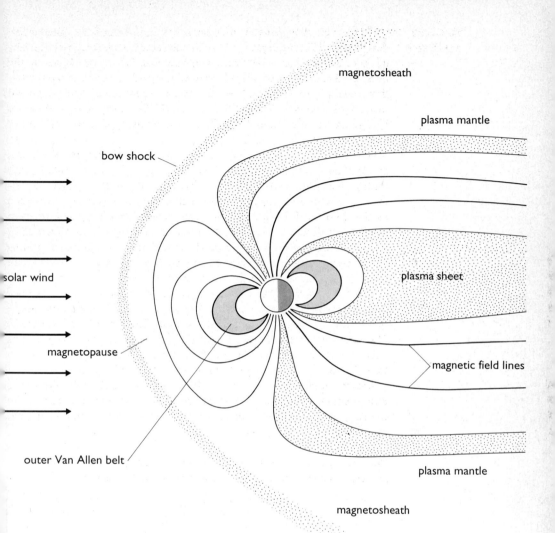

Figure 8.4 *The Earth's magnetosphere, an extended structure embedded in the solar wind. Interactions between the two are complex, and may give rise to spectacular aurorae under certain conditions. Energetic particles from the plasma sheet are accelerated into the high atmosphere, producing light-shows sometimes visible at low latitudes.*

One of the night sky's most spectacular phenomena is the *aurora*, produced in the atmosphere at altitudes above 100 km (60 miles) as a result of collisions between energetic subatomic particles from the Sun, and atoms and molecules of oxygen and nitrogen. These collisions produce excitation, leading to light emission at characteristic wavelengths. The visible light of the aurora is dominated by green (557.7 nm wavelength) and red (630 nm) oxygen emissions, resulting from excitation by electrons accelerated in the Earth's magnetosphere (Figure 8.4) under disturbed conditions.

The aurora

The magnetosphere is the region of space in which the planet's magnetic field has a controlling influence. The Earth's magnetic field results from fluid motions within its liquid metallic core. In isolation, the field would produce a simple dipole pattern, as around a bar magnet. The Earth, however, like all the planets, is embedded in the *solar wind*, a continual outflow of ionized gas (plasma) from the Sun. The solar wind compresses the Earth's magnetic field on the sunward side, and draws it out into the long *magnetotail* downwind. Most of the solar wind is deflected around the Earth – past the *bow shock*, analogous to the pressure wave ahead of a ship cutting through the water. The Earth's magnetosphere is, effectively, a cavity in the solar wind.

The solar wind breezes past the Earth at an average velocity of about 400 km/s (250 mile/s), carrying with it a magnetic field (the *interplanetary magnetic field*) whose strength and orientation are determined by features at the Sun's surface. At times, particularly when sunspot numbers are high, the solar wind can become gusty (up to 2000 km/s, 1250 mile/s) and magnetically turbulent. Under such conditions, the magnetosphere can become disturbed.

At high latitudes there are weak points in the magnetosphere – the polar cusps – which allow a small fraction of solar wind material to penetrate the Earth's magnetic shield on the dayside. Some of this material drizzles directly into the high atmosphere, at fairly low energies, to produce pallid, diffuse aurorae visible only from Arctic or Antarctic locations during the dark midwinter noon.

The aurorae occasionally seen at lower latitudes are produced by an altogether different mechanism. Solar wind plasma entering by the polar cusps on the dayside of the magnetosphere is channelled into the plasma sheet which lies more or less in the Earth's equatorial plane, downwind along the magnetotail. The plasma sheet is a vast reservoir of such particles, and is fairly stable as long as the solar wind remains steady. Under quiet conditions, particles from the plasma sheet circulate along magnetic field lines into the two *auroral ovals*, one around each geomagnetic pole. The auroral ovals normally lie at high latitudes, girdling the polar cap as a fairly narrow ring of auroral activity some 4000 km (2500 miles) in diameter.

When the solar wind turns gusty, however, the situation can change rapidly. The principal source of variation in the solar wind is flare activity, associated with sunspot groups. Flares eject pockets of intense, rapidly fluctuating magnetic field and particles at higher density into the solar wind. If the flare ejection happens to be in the right direction, this energetic pocket arrives in near-Earth space 36–48 hours after the flare, and can wreak havoc with the magnetosphere.

The first effect is a compression of the magnetosphere, momentarily intensifying the Earth's magnetic field. Precise instruments – magnetometers – detect this shock effect as a *sudden storm commencement*. If conditions are favourable (not every flare produces a low-latitude aurora), the subsequent disturbance gives rise to a *geomagnetic storm*, in which the magnetosphere as a whole becomes stressed. Stress in the magnetotail causes particles stored in the plasma sheet to be ejected, both earthwards and down the magnetotail into the solar wind. Plasma sheet particles are accelerated by these events, arriving at high energies in the high atmosphere. During these events the magnetotail acts as a catapult, or giant natural particle accelerator.

During a geomagnetic storm the global magnetic field becomes weakened, and the auroral ovals expand towards the equator. At the same time, the nightside of the oval becomes broader. It is this broadening and expansion towards the equator that carries the aurora into the skies of observers in northwest Europe and the United States.

A geomagnetic storm can last for 24 hours or, more rarely, several days. Disturbances vary in intensity. Small disturbances lead to small auroral oval expansions, and activity remains at fairly high latitudes. Major storms, such as those of 1989 March and 1991 November, can carry the aurora a long way towards the equator, providing a spectacular celestial light-show for observers in the south of England, or even the southern United States.

Observing the aurora
The aurora takes its name from its most common low-latitude appearance – a glow resembling the dawn, low above the poleward horizon (*aurora borealis* means "northern dawn"; the southern hemisphere equivalent is the *aurora australis*). Sadly, such glows are often missed from light-polluted areas, especially if faint. The glow represents the uppermost parts of a display which may be more impressive at higher latitudes. Stars can be seen through it, and foreground clouds may appear silhouetted against it.

Quite often, especially at lower latitudes, the glow fades away without much more being seen. Sometimes, however, a glow appearing in early evening is simply the prelude to better things. As the night progresses, the glow may brighten and rise higher into the sky, becoming unmistakeable. As it rises, the glow may take on a more discrete, recognizable structure, and become an *arc*: a bow-shaped arch of auroral light spanning the poleward sky. Arcs at the early stage of activity are often featureless – *homogeneous*. As activity rises, however, the arc may begin to fluctuate in brightness. Fades may be followed by the development of bright areas along the span of the arc, from which vertical columns – *rays* – develop. Rays often drift slowly along the arc. Sometimes, the only activity seen on a given night will be a few isolated bundles of rays resembling searchlight beams, extending over the poleward horizon.

Folding of an arc produces the ribbon-like form of a *band*. Like an arc, a band may be homogeneous or rayed. Rays vary in length: in some displays they are short, extending 5° or less, while at other times they may extend for long distances, even past the zenith. Rayed bands in which the rays stretch towards the zenith give rise to the curtain or drapery effect which is popularly portrayed as typical of the aurora.

Rayed aurora can sometimes be colourful. During active displays, the base of a rayed arc or band appears bluish or green, while the upper parts are predominantly reddish. These colour contrasts result from the variation of the atmosphere's density with altitude. Faint aurorae often appear colourless to the eye, but may register as greenish on long-exposure colour photographs.

During the most extreme disturbances, an aurora will fill half the sky, and more, with rays – not ideal conditions, it must be admitted, for meteor observing. At the very peak of a storm, and rather seldom at lower latitudes, the display can pass overhead and into the equatorward half of the sky. At this stage, perspective comes into play, just as it does with the meteor radiant effect, so that auroral rays and other features appear to converge on a single area of sky, producing the form of a *corona*. The centre of the corona is in the *magnetic zenith* – the point towards which a free-swinging dip needle would swing, some 10–15° equator-wards of the true zenith for mid-latitude observers.

The corona at low latitudes is usually fairly short-lived, and the display soon falls back polewards. Rarely – as during the great aurora of 1989 March 13–14 – the corona may persist for much of the night, depending on the observer's latitude.

After a retreat polewards, the display may subside to a glow and fade away, but on other occasions activity may build up again, and several peaks may be seen in a single night. One theory is that recurrent peaks reflect the time taken for stress on the magnetotail to build up again – reloading the catapult, as it were.

Figure 8.5 *Useful altitude and azimuth measurements for auroral features.*

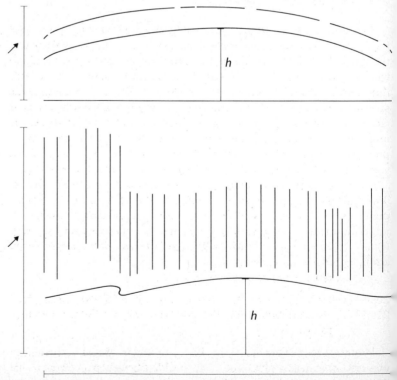

azimuth

Figure 8.6 *The spectacular aurora of 1991 November 8–9, photographed at 0113 UT from Chichester by the author.*

Aurorae can still be usefully observed by amateur astronomers, despite the advent of satellite monitoring. Positional measurements help define the geographical spread of activity, though an observer at a given location will see only a small section of the auroral oval. Useful observation records must include an indication of the auroral forms – glow (G), homogeneous arc (HA), rayed arc (RA), homogeneous band (HB), or rayed band (RB) – present at a particular time. The length of any rays present is indicated by a subscript from 1 to 3, from shortest to longest, and a prefix is added to indicate the state of auroral activity – quiet (e.g. qHA), active (e.g. aR_1B), multiple (e.g. mR_2R), or even coronal (e.g. acR_3B).

It is important to estimate the display's extent in altitude and azimuth. Two altitude measurements are useful: h, the altitude of the highest point on the *base* of an arc or band, and ↗, indicating the display's uppermost reaches (Figure 8.5). Also note the brightness, ranging from (i) faint, comparable to the Milky Way, to (iv) very bright, possibly even strong enough to cast shadows.

Typical records might therefore read:

2030 UT: qHA brightness (i), 320–030° azimuth, h 10°↗15°. Pale white.
2115 UT: aR_2B brightness (ii), 300–040° azimuth, h 15°↗45°. Red tops
        to rays.

Photographs of aurorae can be spectacular. Most experienced observers favour fast (ISO 400) colour slide films such as Ektachrome 400 or Fujichrome 400, as they give good colour reproduction, and are less prone to the garish "technicolour" responses which afflicted many earlier emulsions. A 28 mm wide-angle lens is excellent for aurora work, preferably used at $f/2.8$ or faster. Exposures range from 30–60 seconds for very faint aurorae to 5–10 seconds for the very brightest. If the aurora is rapidly moving, exposures should be kept short to avoid loss of detail. Figure 8.6 shows the rayed arc of the spectacular 1991 November 8–9 display, as I recorded it from Chichester at 0113 UT.

The appearance of aurorae at lower latitudes is, to a large degree, unpredictable. In general, the chances of seeing an aurora are improved by high levels of sunspot activity, but even when solar flares have been recorded by observers monitoring the Sun in hydrogen-alpha light, there is no guarantee that enhanced geomagnetic activity will follow. Coupling between the interplanetary and terrestrial magnetic fields depends largely on the orientation of the former: only flares producing a strong southerly magnetic component are efficient in generating magnetic storms at the Earth. This confounds attempts to forecast auroral activity, and major alerts issued in 1989 June, and again in 1991, proved to be false alarms. The best advice I can offer anyone hoping to see an aurora is, simply, to keep an eye on the the sky – and this, of course, is where regular meteor observers who spend long hours outdoors have a distinct advantage. Many of the reported sightings of aurorae come from meteor watchers who happened to be out observing when activity became apparent.

Noctilucent clouds

The ultimate fate of most meteoroids entering the Earth's atmosphere is disintegration, through the processes discussed in Chapter 2. Following the break-up of the incoming body, small fragments are left suspended in the high atmosphere, from where they gradually settle out as micrometeorites or Brownlee particles. Such debris is also thought by many to provide the nuclei around which *noctilucent clouds* (NLCs) can condense.

NLCs are seen from latitudes above 50° during the short summer nights, at times when the Sun is between 6° and 16° below the poleward horizon. At higher latitudes, such conditions prevail throughout the night. Most observations are made from the northern hemisphere, where the main populated land-masses are at latitudes which coincide more closely with the zones in which NLCs occur than in the southern hemisphere. Observers in Canada and northwest Europe have accumulated large numbers of useful reports of the phenomenon since the 1950s.

The clouds form when small traces of water vapour, carried aloft by atmospheric upwelling in the spring, reach cooler altitudes in the summer. Under favourable conditions, the water vapour then condenses onto small nuclei (most likely provided by meteoric debris) to produce extremely tenuous clouds whose peak altitude lies around 82 km (51 miles) above the Earth, five times higher than the highest "weather" clouds – the familiar mare's-tails of cirrus, to which NLCs bear a superficial resemblance. NLCs are so tenuous that they cannot be seen during daylight as the brightness of the background sky drowns them out. Only during the twilit conditions of late summer evenings can they be seen, still sunlit against the darkening sky, after any foreground tropospheric clouds have been immersed in the Earth's advancing shadow. It is from this "night-shining" nature that NLCs take their name.

They form near the mesopause, a layer of the atmosphere not far below the heights at which aurorae appear. It has long been assumed that there is a negative correlation between the occurrence of aurorae and NLCs: frequent auroral activity is thought to raise temperatures in the mesopause to a level at which NLCs are less likely to form, around sunspot maximum. However, some doubt has been cast on this by Canadian and Scottish observations of aurorae and NLCs occurring together.

It has also been speculated that NLCs are somehow related to the daytime meteor activity of the Beta Taurids. In the northern hemisphere, NLCs may peak around late June each year, near the maximum of the meteor shower. Long-term studies, however, suggest that the negative correlation between high geomagnetic activity and NLC occurrence is stronger – the principal controlling influence is the temperature of the mesopause.

One thing does seem certain: sightings of NLCs from northwest Europe almost doubled in frequency between the 1950s and the late 1980s. A possible explanation is that industrial pollution may be adding to the water vapour content of the high atmosphere, making it easier for NLCs to form. Further observations are required to see whether this increase in NLC frequency persists into the early twenty-first century.

Observing noctilucent clouds

Many observers have probably seen NLCs without realizing exactly what they were. In a good year, particularly around the optimum observing latitude of 56–57°, NLCs may be present on one night in three, or even more often. The clouds have a distinctive silvery-blue colour, shading off to gold towards the horizon, which, once seen, is not forgotten.

Figure 8.7 *Sketches of typical noctilucent cloud forms. Type I NLC, not shown here, is a featureless veil which may appear as a background to other structures. Type II, consisting of parallel bands of cloud, is fairly common. Interwoven structure is described as Type III, while curved forms, or "billows," are Type IV. Sometimes dense patches with no apparent internal structure are seen; this is Type V NLC.*

Type II
Type IV
Type III
Type V

Like aurorae, NLCs can be observed with the naked eye, but it is sometimes worth using binoculars as an aid to positive identification. In binoculars, they show a delicate, sharp interwoven structure, quite unlike cirrus. Tropospheric clouds appear much less distinct than NLCs when viewed under binocular magnification in the summer twilight. The clouds can take one of several forms, which should be identified in observational reports. These are shown schematically in Figure 8.7. When it occurs alone, Type I NLC is probably quite frequently missed.

Displays of NLCs vary in extent and brightness. From lower latitudes they are usually confined to the northern sky below Capella, having a maximum elevation of about 10–15°, but from Scottish or Canadian latitudes displays can be more extensive: from Edinburgh I have seen displays which covered more than half the sky. While the particles which make up NLCs are difficult to examine directly, one of their known properties is that they scatter sunlight predominantly forwards. This means that NLCs in the direction of the Sun are brighter than those overhead or to the south. The tenuous nature of the clouds adds to the difficulty of seeing them overhead.

The extent and brightness of NLC displays varies gradually during the night, depending on how far the Sun is below the northern horizon. Displays are often at their most extensive – but are also rather over-whelmed by the background sky brightness – just after sunset, and again just before sunrise. Towards midnight the illuminated portion of the NLC field sinks lower towards the northern horizon; at lower latitudes, such as those of southern England, the clouds may fade from view altogether as even they become immersed in the Earth's shadow for a time. The brightest region in an NLC display is often just above the Sun's position, and so gradually moves around the horizon from northwest to northeast as the night wears on.

Unlike active aurorae, NLC displays change in appearance relatively slowly. Many changes are simply a result of the gradually altering solar illumination of the clouds, but movement is apparent in some displays, indicative of high-atmosphere winds blowing east to west at speeds of up to 400 km/h (250 mile/h); these winds are also responsible for distorting long-duration meteor trains, as discussed in Chapter 7.

Reports of visual sightings of NLCs are of interest to professional scientists studying the upper atmosphere, and are collected in northwest Europe by the BAA Aurora Section. A similar network is coordinated in North America by Mark Zalcik.

Since they change only slowly, it is normally sufficient to record the appearance of NLCs at 15-minute intervals – preferably on the hour, then at 15, 30, and 45 minutes past the hour. Reports should include the type, or types, of NLC present at a particular time (UT), and their extent in altitude and azimuth – as for aurorae. Brightness can be indicated, on a scale from (i) faint to (iii) very bright. Annotated sketches can be used to get the details down quickly. It can also be useful to make reports of those clear nights when there was definitely no NLC visible from a given location.

Figure 8.8 *A bright noctilucent cloud display photographed from Edinburgh on 1983 June 30–31.*

As with aurorae, NLCs present considerable scope for photography. Results are often very pleasing, since features on the horizon almost inevitably appear in the photographs. The background light level during the late evening when NLCs become visible is usually sufficient to ensure that both landscape and clouds add to the aesthetic appeal. The photographs will show few stars, however, since usually only Capella and Beta Aurigae are bright enough to cut through the twilight low to the north.

Colour film does most justice to NLCs, and my own preference is for Ektachrome 400 slide film. Depending on the display's extent, either a "standard" 50 mm or wide-angle 28 mm lens can be used, at $f/2.8$. Exposure times at ISO 400 depend on the brightness of the background sky, and of the NLCs, and can be from 1 to 3 seconds. Exposures should be kept short when the sky is bright, since overexposure will result in the background drowning out the NLCs in the eventual image. Longer exposures may be necessary as the display fades in the darkening sky around midnight. Pairs of precisely timed photographs taken from well separated locations have been used to triangulate NLCs. Figure 8.8 shows a fine display which I photographed from Edinburgh on 1983 June 30–31.

In northwest Europe and Canada, NLCs are seen in a 10-week "season" from late May to early August. Observing them is a useful activity for those summer nights at higher latitudes when the sky remains too bright for meteor watches.

Bibliography

Beatty, J.K., and Chaikin, A. (editors), *The New Solar System*, 3rd edition, Cambridge University Press/Sky Publishing Corp., 1990.

Bone, N., *The Aurora: Sun–Earth Interactions*, Ellis Horwood, 1991.

Burke, J.G., *Cosmic Debris: Meteorites in History*, University of California Press, 1986.

Duffett-Smith, P., *Practical Astronomy with Your Calculator*, 3rd edition, Cambridge University Press, 1988.

Kronk, G.W., *Comets: A Descriptive Catalog*, Enslow, 1984.

Kronk, G.W., *Meteor Showers: A Descriptive Catalog*, Enslow, 1988.

Lovell, B., *Astronomer by Chance*, Oxford University Press/Basic Books, 1990.

McDonnell, J.A.M. (editor), *Cosmic Dust*, Wiley, 1978.

Moore, P., and Mason, J., *The Return of Halley's Comet*, Patrick Stephens/W. W. Norton, 1984.

Ridpath, I. (editor), *Norton's 2000.0*, Longman/Wiley, 1989.

Roggemans, P. (editor), *Handbook for Visual Meteor Observations*, Sky Publishing Corp., 1989.

Tirion, W., Rappaport, B., and Lovi, G., *Uranometria 2000.0* (2 volumes), Willmann-Bell, 1988.

Whipple, F.L., *The Mystery of Comets*, Smithsonian Institution, 1985.

Specialist articles

The magazines *Astronomy Now* (UK) and *Sky & Telescope* and *Astronomy* (US) often carry notes on recent or forthcoming meteor activity. More detailed papers are usually to be found in publications such as the *Journal of the British Astronomical Association*. In the next column is a selection of articles covering observational results, which give something of the flavour of amateur meteor studies.

"Meteors by the minute," *Sky & Telescope*, Vol. 60, pp. 446–447, 1980.

"The 1981 Perseids: Fine indeed," *Sky & Telescope*, Vol. 62, pp. 624–626, 1981.

Bone, N.M., "Visual observations of the Taurid meteor shower 1981–1988," *Journal of the British Astronomical Association*, Vol. 101, pp. 145–152, 1991.

Bone, N., and Evans, S., "The Perseid meteor shower in 1988," *Astronomy Now*, Vol. 2, No.11, pp. 51–53, 1988.

di Cicco, D., "The 1991 Perseids: Picking up!" *Sky & Telescope*, Vol. 83, pp. 225–227, 1992.

Evans, S., "Photographic observations of the Perseid meteor stream in 1988," *Journal of the British Astronomical Association*, Vol. 100, pp. 169–172, 1990.

Hindley, K.B., "The Quadrantid meteor stream," *Sky & Telescope*, Vol. 43, pp. 162–164.

Lynch, J.L., "A different way to observe the Perseids," *Sky & Telescope*, Vol. 84, pp. 222–225, 1992. (An introduction to radio observation.)

Mason, J.W., and Sharp, I.D., "The Perseid meteor stream in 1980," *Journal of the British Astronomical Association*, Vol. 91, pp. 368–390, 1981.

Spalding, G.H., "The Geminid meteor stream in 1980," *Journal of the British Astronomical Association*, Vol. 92, pp. 227–235, 1982.

Spalding, G.H., "The Giacobinid meteor shower – Prospectus for 1985," *Journal of the British Astronomical Association*, Vol. 95, pp. 211–212, 1985.

Spalding, G.H., "The activity of the Orionid meteor stream in 1985," *Journal of the British Astronomical Association*, Vol. 98, pp. 26–34, 1987.

Useful addresses

Several bodies around the world coordinate amateur meteor observations. Of these, the longest established are probably the British Astronomical Association's Meteor Section, and the American Meteor Society. In the UK, the Junior Astronomical Society, which caters for beginners of all ages, has a thriving Meteor Section. An International Meteor Organization, centred mainly in the Benelux countries of Europe, has operated since 1988.

There may be a local astronomical society or club near you. Such organizations are an excellent source of useful contacts, information, and advice, and membership is strongly recommended.

British Astronomical Association
Burlington House
Piccadilly
London W1V 9AG
England

Junior Astronomical Society
36 Fairway
Keyworth
Nottingham NG12 5DU
England

American Meteor Society
Department of Physics and
   Astronomy
State University of New York at
   Geneseo
Geneseo
NY 14454
USA

Association of Lunar and
   Planetary Observers
3930 Raven Drive
Waco
TX 76712
USA

International Meteor
   Organization
Pijnboomstraat 25
B–2800 Mechelen
Belgium

Mark Zalcik
#2 14225–82 Street
Edmonton
Alberta
Canada T5E 2V7

Radio Society of Great Britain
Lambda House
Cranborne Road
Potters Bar
Hertfordshire EN6 3JE
England

American Radio Relay League
Newington
CT 06111
USA

Index